AGRICULTURE ISSUES AND POLICIES

PROFITABILITY OF ORGANIC FIELD CROPS

AGRICULTURE ISSUES AND POLICIES

Additional books in this series can be found on Nova's website under the Series tab.

Additional e-books in this series can be found on Nova's website under the e-book tab.

AGRICULTURE ISSUES AND POLICIES

PROFITABILITY OF ORGANIC FIELD CROPS

MADELINE ROSE BOWERS
EDITOR

New York

Copyright © 2015 by Nova Science Publishers, Inc.

All rights reserved. No part of this book may be reproduced, stored in a retrieval system or transmitted in any form or by any means: electronic, electrostatic, magnetic, tape, mechanical photocopying, recording or otherwise without the written permission of the Publisher.

We have partnered with Copyright Clearance Center to make it easy for you to obtain permissions to reuse content from this publication. Simply navigate to this publication's page on Nova's website and locate the "Get Permission" button below the title description. This button is linked directly to the title's permission page on copyright.com. Alternatively, you can visit copyright.com and search by title, ISBN, or ISSN.

For further questions about using the service on copyright.com, please contact:
Copyright Clearance Center
Phone: +1-(978) 750-8400   Fax: +1-(978) 750-4470   E-mail: info@copyright.com

NOTICE TO THE READER

The Publisher has taken reasonable care in the preparation of this book, but makes no expressed or implied warranty of any kind and assumes no responsibility for any errors or omissions. No liability is assumed for incidental or consequential damages in connection with or arising out of information contained in this book. The Publisher shall not be liable for any special, consequential, or exemplary damages resulting, in whole or in part, from the readers' use of, or reliance upon, this material. Any parts of this book based on government reports are so indicated and copyright is claimed for those parts to the extent applicable to compilations of such works.

Independent verification should be sought for any data, advice or recommendations contained in this book. In addition, no responsibility is assumed by the publisher for any injury and/or damage to persons or property arising from any methods, products, instructions, ideas or otherwise contained in this publication.

This publication is designed to provide accurate and authoritative information with regard to the subject matter covered herein. It is sold with the clear understanding that the Publisher is not engaged in rendering legal or any other professional services. If legal or any other expert assistance is required, the services of a competent person should be sought. FROM A DECLARATION OF PARTICIPANTS JOINTLY ADOPTED BY A COMMITTEE OF THE AMERICAN BAR ASSOCIATION AND A COMMITTEE OF PUBLISHERS.

Additional color graphics may be available in the e-book version of this book.

Library of Congress Cataloging-in-Publication Data

ISBN: 978-1-63484-167-2

Published by Nova Science Publishers, Inc. † New York

CONTENTS

Preface		**vii**
Chapter 1	The Profit Potential of Certified Organic Field Crop Production *William D. McBride, Catherine Greene, Linda Foreman and Mir Ali*	**1**
Chapter 2	Organic Farming Systems *Catherine Green and Robert Ebel*	**55**
Index		**61**

Preface

Organic crop acres in the United States more than doubled between 2002 and 2011 as acreage increased from 1.3 to over 3 million acres. While acreage for some major field crops increased substantially during this period, growth was more modest or had stalled for others. This book examines the profitability of corn, wheat, and soybean production using national survey data and finds that significant economic returns are possible from organic production of these crops. The main reason for higher per-bushel returns to organic production is the price premiums paid for organic crops. Despite potentially higher returns, the adoption of organic field crop production has been slow and is challenging due to such factors as achieving effective weed control and the processes involved with organic certification.

In: Profitability of Organic Field Crops ISBN: 978-1-63484-167-2
Editor: Madeline Rose Bowers © 2015 Nova Science Publishers, Inc.

Chapter 1

THE PROFIT POTENTIAL OF CERTIFIED ORGANIC FIELD CROP PRODUCTION[*]

William D. McBride, Catherine Greene, Linda Foreman and Mir Ali

ABSTRACT

Organic crop acres in the United States more than doubled between 2002 and 2011 as acreage increased from 1.3 to over 3 million acres. While acreage for some major field crops increased substantially during this period, growth was more modest or had stalled for others. This study examines the profitability of corn, wheat, and soybean production using national survey data and finds that significant economic returns are possible from organic production of these crops. The main reason for higher per-bushel returns to organic production is the price premiums paid for organic crops. Despite potentially higher returns, the adoption of organic field crop production has been slow and is challenging due to such factors as achieving effective weed control and the processes involved with organic certification.

---

[*] This is an edited, reformatted and augmented version of Economic Research Report Number 188, issued by the U.S. Department of Agriculture, Economic Research Service, July 2015.

WHAT IS THE ISSUE?

Certified organic crop acres more than doubled between 2002 and 2011, as acreage increased from 1.3 million acres to over 3 million acres. A large part of this growth was in major field crops—corn, soybeans, and wheat—where certified organic production increased about 264,000 acres. Despite this interest in organic agriculture and its potential to address environmental concerns, little information is available about the relative costs and returns of organic grain production on commercial farms. Most previous research is derived from results of long-term experimental field trials and offers limited economic analysis. Results of this study provide information about potential economic returns from organic field crop production on commercial farms and the additional costs incurred from producing organic.

WHAT DID THE STUDY FIND?

This study of field crop production indicates a profit potential from organic systems that is primarily due to the significant price premiums paid for certified organic crops. Additional economic costs of organic versus conventional production were more than offset, on average, by higher returns from organic systems for corn and soybeans, although not for wheat. Other findings of this study:

- Organic field crop production was, on average, conducted on farms with less total acreage and less field crop acreage than conventional farms. Despite having fewer acres, producers of some organic field crops were less likely to work off-farm. These producers were also more likely to have attended college than conventional producers. Organic production more often occurred in northern States where pest pressures are less severe.
- Production practices used on organic and conventional field crop operations were quite different. Most conventional producers of corn and soybeans used genetically modified seed varieties not allowed for certified organic crop production. Most organic producers used mechanical practices, such as tillage and cultivating for weed control, while conventional producers rarely used a cultivator and relied mainly on chemical weed control. Organic corn and soybean

producers more often rotated row crops with small grain and meadow crops and often included an idle year in the rotation. Conventional producers of these crops mainly used a rotation consisting of continuous row crops.
- Much of the experimental research on organic field crop production has found similar yields and lower per-acre costs from organic relative to conventional field crop production.

However, the economic analysis used with the experimental research has primarily examined only operating or variable costs, excluding the economic costs of such resources as land, labor, and capital. Findings of this observational study of commercial organic and conventional field crop production found lower yields and mostly higher per-acre total economic costs from organic systems.

- As in much of the economic analyses using experimental data, per-bushel operating costs of organic relative to conventional systems were similar in this study. However, the per-bushel economic costs of organic production were significantly higher because of the higher per-acre costs and lower yields.
- The economic costs of organic compared with conventional production estimated in this study were roughly between $83 and $98 per acre higher for corn, $55-$62 per acre higher for wheat, and $106-$125 per acre higher for soybeans. These estimated cost differences are all higher than those suggested by the relative means.
- Results of this study imply that some conventional farms may be able to earn greater returns if transitioned to organic production. Nevertheless, adoption of the organic approach among U.S. field crop producers remains extremely low. Perhaps a key factor is that organic field crop production is particularly challenging compared with conventional production in achieving effective weed control and crop yields. Also, the processes involved with organic certification can be complex and time-consuming.

HOW WAS THE STUDY CONDUCTED?

The profitability of organic field crop production was examined using Agricultural Resource Management Survey (ARMS) data from corn, wheat, and soybean producers that included targeted samples from organic growers. Two procedures were used to calculate the difference between conventional and organic crop production costs:

1) Propensity-score matching generated a sample of similar conventional and organic producers of each crop based on observed farm and operator characteristics from which to measure the difference in organic and conventional production costs.
2) Regression with endogenous treatment-effects was employed to describe this same difference in organic and conventional production costs, accounting for the impact of both observable and unobservable variables on crop production costs.

Results of these procedures were compared with the difference in mean cost-of-production estimates for organic and conventional producers. Estimated organic transition and certification costs were added to each result, and the cost differences between organic and conventional crop production systems were compared with historic price premiums paid for organic crops to evaluate the potential profitability of organic field crop production. Despite the detailed producer survey data used in this study, the limited time-series data dimension renders this study primarily one of association rather than causality.

INTRODUCTION

Organic cropping systems rely on ecologically based practices, such as biological pest management and composting, and exclude most synthetic chemicals. Under organic cropping systems, the fundamental components and natural processes of ecosystems—such as soil organism activities, nutrient cycling, and species distribution and competition—are used as farm management tools (Greene and Kremen, 2003). For example, crops are rotated, pest prevention techniques are employed, animal manure and crop residues are recycled, and planting/harvesting dates are carefully managed. Major reasons for the popularity of organic farming are the low impact on the environment; the ability to farm without relying on a limited resource, synthetic nitrogen, which has negative environmental consequences such as nitrate pollution of groundwater and waterways; and the perception that organic food is more healthful. While economic concerns are important, they are not always the main reason farmers choose the organic approach.

"Certified organic" is a labeling term that indicates that the food or other agricultural product has been produced through approved methods that integrate cultural, biological, and mechanical practices that foster cycling of resources, promote ecological balance, and conserve biodiversity (USDA/AMS a). In the United States, the National Organic Program (NOP) is the Federal regulatory framework governing organic food and also is the name of the organization within the U.S. Department of Agriculture (USDA) responsible for administering and enforcing the regulatory framework. The Organic Foods Production Act of 1990 required that the USDA develop national standards for organic products. The NOP final rule was published in the *Federal Register* in December 2000 (*Federal Register*, 2000).

The Organic Foods Production Act of 1990 "requires the Secretary of Agriculture to establish a National List of Allowed and Prohibited Substances, which identifies synthetic substances that may be used and the nonsynthetic substances that cannot be used in organic production and handling operations." USDA promulgated regulations establishing the NOP standards and the USDA program in 2000. Certification is handled by State, nonprofit, and private agencies that have been approved by USDA. Under the NOP, farmers who wish to use the word "organic" in reference to their business and products must be certified organic.[1] In addition to restrictions on which substances may be used to qualify for organic certification, certain production practices, such as crop rotations and pasture feeding requirements for ruminant animals, must be followed in order to maintain the organic certification status.

BACKGROUND AND OBJECTIVE

U.S. crop acres under USDA certified organic systems have grown rapidly since the NOP was implemented in 2002. Organic crop acres were nearly 2.5 times higher in 2011 than in 2002, as acreage increased from about 1.3 million to almost 3.1 million acres (USDA/ERS a). While acreage for some major field crops increased substantially during this period, growth was more modest or had stalled for others. Among the three major field crops examined in this study—corn, soybeans, and wheat—certified organic production of corn increased the most, from about 96,000 acres in 2002 to 131,000 acres in 2005, to 234,000 acres in 2011 (fig. 1). Certified organic soybean acreage declined from a peak of 175,000 acres in 2001 to 100,000 acres in 2007, but rebounded to 132,000 acres in 2011. Organic wheat acreage was the largest in all years, starting from 225,000 acres in 2002, increasing to 294,000 acres in 2005 and peaking at more than 400,000 acres in 2008, before falling to 345,000 acres in 2011.

Much of the increased organic corn production has been to support a rapidly growing organic dairy sector in which the number of certified organic milk cows increased nearly fourfold from about 67,000 in 2002 to nearly 255,000 in 2011 (USDA/ERS a). Higher prices for conventional corn, soybeans, and wheat since 2008 and somewhat slower demand growth for organic products due to the economic recession, along with increasing imports of these crops, may have helped limit increases in U.S. organic acreage in more recent years (USDA/NASS a).

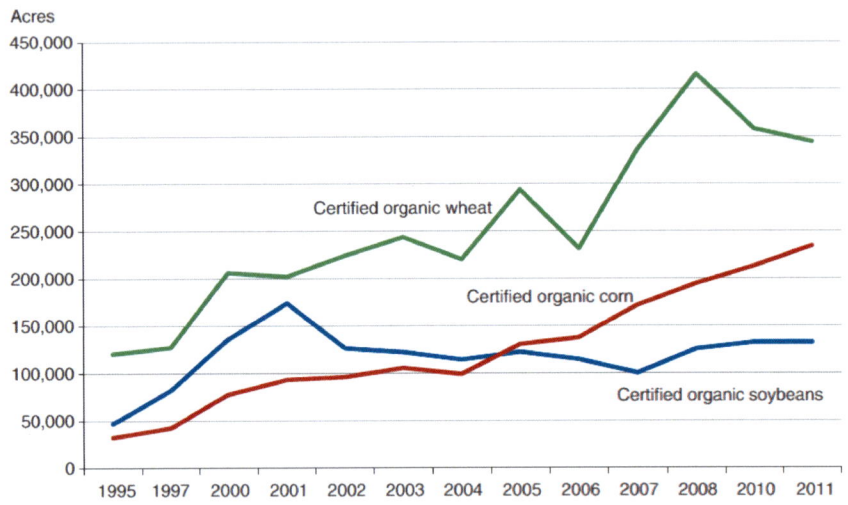

Figure 1. U.S. organic corn, wheat, and soybean acreage, 1995-2011.[1]

[1] Organic crop acreage data were not available for 1996, 1998, 1999, and 2009.
Source: USDA, Economic Research Service.

Organic production is facilitated in the United States through a cost-share program offered by USDA consisting of the National Organic Certification Cost Share Program (NOCCSP) and the Agricultural Management Assistance (AMA) Organic Certification Cost Share Program (USDA/AMS b). These programs help to defray the cost of organic certification by authorizing USDA to allocate funds under the NOCCSP and AMA to eligible State agencies. The State agencies then reimburse certified organic operators for a portion of the costs the operators incur to obtain or maintain organic certification. In 2015, individual operators were eligible for reimbursement of 75 percent of their certification costs up to a maximum of $750 per category of certification. Total funding for these programs in 2015 was set at $10.3 million for the NOCCSP and $900,000 for the AMA (USDA/AMS a).[2]

Despite the interest and support of organic crop production in the United States, overall adoption of organic corn, soybeans, and wheat remains low, standing at less than 1 percent of the total acreage of each crop in 2011 (USDA/NASS b). Low levels of organic adoption among U.S. field crop producers may be affected by the dearth of information about the relative costs and returns of organic and conventional production systems on commercial farms in the United States and the performance of farms that are choosing the organic approach. Several researchers (Delate et al., 2003.; Mahoney et al., 2001; Hanson et al., 1997; Pimentel et al., 2005; Smith et al., 2004; among others) have examined organic crop production in a long-term experimental setting, but little has been reported about the commercial production of organic field crops (McBride and Greene, 2009; Nordquist et al., 2014). This study utilizes observational data obtained in samples of U.S. field crop producers from USDA's Agricultural Resource Management Survey (ARMS) in a comparison of conventional and organic systems. The main objective is to estimate the difference in costs of production that can be attributed to producing certified organic crops, using these costs to indicate price premiums that make organic systems profitable when compared with conventional systems (see "Data" box).

Data from samples of U.S. corn, wheat, and soybean producers, including targeted samples of certified organic producers of each crop as part of the 2010, 2009, and 2006 ARMS, respectively, support the research in this study. (In this context, corn refers to field corn, excluding such specialty crops as sweet corn and popcorn.) This study contrasts the costs of organic and conventional production using two distinct empirical procedures commonly used in the literature to evaluate relationships in observational data. Both procedures use a treatment-effect analysis where the treatment is organic production and its effect is examined on different levels of production costs. First, a matched sample of organic and conventional producers, based on farm, operator, and production characteristics, was generated in order to account for selection-bias in measuring the organic treatment-effect on production costs. This is referred to as "propensity-score matching." Second, a "regression with endogenous treatment-effects" was conducted to account for observable differences between organic and conventional crop producers and potential unobservable differences resulting from selection-bias in the assignment of organic production among producers in the population (see "Appendix: Empirical Procedure").

The two treatment-effect measures were compared with the mean difference in the production costs of organic and conventional producers. This comparison indicates whether the mean difference of costs is misleading, as would be expected, in a situation where producers self-select the treatment. Organic transition and certification costs were then added to the estimated differences in costs to account for these additional costs of certified organic production. The estimated cost differences were then compared with historic organic price premiums.

Data

Data used in this study come from USDA's 2010, 2009, and 2006 Agricultural Resource Management Survey (ARMS) administered by the National Agricultural Statistics Service (NASS) and Economic Research Service (ERS). The ARMS data include farm financial information, such as farm income, expenses, assets, and debt, as well as farm and operator characteristics. This study uses ARMS versions that include information about the production practices and costs of U.S. commodity production—corn in 2010, wheat in 2009, and soybeans in 2006. Each version targeted producers in States that included over 90 percent of U.S. planted acreage of the crop in each year.

The 2010 ARMS corn sample consisted of 3,893 farms with 627 samples targeting organic operations. After accounting for out-of-business operations, survey refusals, and questionnaires with incomplete data, 1,087 conventional corn farms and 243 organic corn farms from Illinois, Indiana, Iowa, Kansas, Michigan, Minnesota, Missouri, Nebraska, New York, North Dakota, Ohio, Pennsylvania, South Dakota, and Wisconsin were used in this study. Of the total 2009 ARMS wheat sample of 3,699 farms, 483 samples targeted organic operations. After accounting for nonresponse and incomplete data, 1,339 conventional wheat farms and 182 organic wheat farms from Colorado, Idaho, Illinois, Kansas, Michigan, Minnesota, Missouri, Nebraska, North Dakota, Ohio, Oklahoma, Oregon, South Dakota, Texas, and Washington were used. The 2006 ARMS soybean sample included 4,557 farms, 907 samples targeted organic operations in 15 States. Of these, 2,209 farms were available for analysis, including 238 operations producing organic soybeans.

> Characteristics and production costs were compared among conventional and organic soybean producers in Illinois, Indiana, Iowa, Kansas, Michigan, Minnesota, Missouri, Nebraska, North Dakota, Ohio, South Dakota, and Wisconsin including 1,425 conventional and 237 organic producers. Farm survey weights on the ARMS data ensure that samples expand to represent the appropriate crop acreage in the surveyed States, and that organic operations represent their correct proportion of the target population despite their disproportionate share of the sample.

CHALLENGES OF ORGANIC FIELD CROP PRODUCTION

Organic field crop producers were asked in the ARMS what they considered to be the most difficult aspect of organic crop production. Categories presented for the choosing by corn and wheat producers were identical and offer insights into the major issues faced by organic field crop producers. The question posed to organic soybean producers did not have the detailed categories presented to the other crop producers and are not directly comparable. More than 40 percent of organic soybean producers reported that achieving yields was the most difficult aspect of organic crop production, also rated highly by corn and wheat producers. The second highest category listed by soybean producers was "other," reported by almost 38 percent of producers. Because a high percentage of soybean producers reported "other," the categories were refined for the subsequent wheat and corn surveys.

The top three aspects of organic production reported as most difficult by corn and wheat producers, and the percent of producers reporting each aspect, are shown in figure 2. Controlling weeds was reported most often, by about 40 percent of both corn and wheat producers. Limited chemical-input options available to organic crop producers makes weed control more difficult in many instances relative to conventional production. Achieving desired crop yields was reported by about 17 percent of organic wheat producers and 12 percent of organic corn producers. Issues associated with achieving organic yields comparable with conventional yields are related to controlling weeds in organic fields, but could also be associated with the performance of NOP-approved organic seed varieties versus conventional seed varieties, and with organic fertilizer options. The soil health of organically managed acreage, typically measured by soil carbon, has also been emphasized as a critical factor affecting the yields from organic relative to conventional field crop acreage (Cavigelli et al., 2013; Coulter et al., 2013).

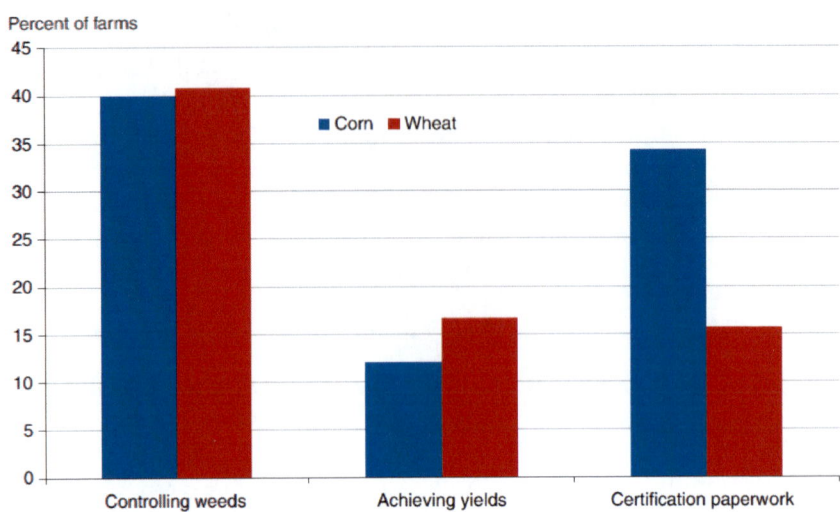

Source: USDA, Economic Research Service and USDA, National Agricultural Statistics Service, 2009 and 2010 Agricultural Resource Management Survey.

Figure 2. Most difficult aspects of organic corn (2010) and wheat (2009) production.

Organic certification paperwork was reported as the most difficult aspect of organic production by 17 percent of wheat producers, and over a third of corn producers. Certification paperwork may be more arduous and time-consuming for corn relative to wheat because input use for corn is much greater than for wheat and plans to meet certified organic requirements may be more complex. Certification paperwork was not listed as an option for soybean producers and may account for the high percentage of these producers reporting other as the most difficult aspect of organic soybean production.

ORGANIC AND CONVENTIONAL FIELD CROP PRICES

Differences in organic and conventional crop production costs provide only part of the information that determines the profitability of organic production systems. Greater prices received for organic field crops can offset higher organic production costs.

Average monthly conventional corn prices rose from less than $5 per bushel starting in 2011 to more than $7 per bushel in 2012 before again falling under $4 per bushel during the second half of 2014 (fig. 3). During 2011-14, organic corn prices followed a similar pattern as conventional corn prices but with wider fluctuations. The gap between average monthly conventional and organic corn prices rose steadily from 2011 through 2012 with organic feed and food corn reaching above $16 per bushel, to reach about $9-$10 higher than conventional corn prices by the end of 2012. Organic corn prices declined in 2013, but the price differential between organic and conventional corn remained in the $5 to $7 per bushel range. While conventional corn prices continued to fall in 2014, organic corn prices increased to around $14 per bushel before falling later in 2014 to around $12. During 2014 the price differential was in the $8-$10 per-bushel range.

As with corn, the price premium for organic wheat generally widened during the 2011 to 2014 period, but the gap between average monthly conventional and organic wheat prices varied greatly for food- and feed-grade wheat. Between 2011 and 2013, price premiums for organic food wheat were generally higher than for organic feed wheat by about $2-$6 per bushel, but widened during 2014 as organic food wheat prices rose to about $18 to $20 per bushel (fig. 4). Farm prices received for organic feed wheat varied significantly during 2011-13, much of the time only $1 to $4 per bushel higher than those for conventional wheat, but also widened in 2014 (fig. 5). Conventional wheat prices were more stable over this period, generally ranging between $6.00 and $8.50 per bushel.

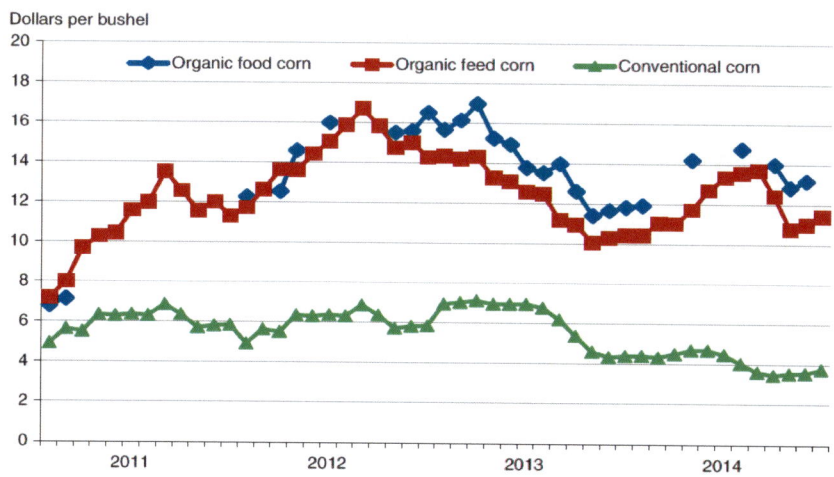

Source: Organic prices from USDA, Agricultural Marketing Service; conventional prices from USDA, National Agricultural Statistics Service.

Figure 3. Organic food and feed, and conventional corn prices, 2011-14.

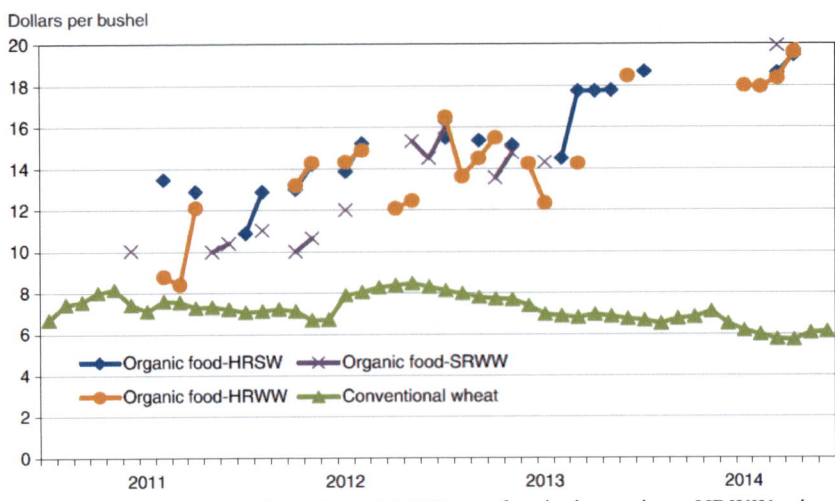

Note: HRSW = hard red spring wheat, SRWW = soft red winter wheat, HRWW = hard red winter wheat.

Source: Organic prices from USDA, Agricultural Marketing Service; conventional prices from USDA, National Agricultural Statistics Service.

Figure 4. Organic food and conventional wheat prices, 2011-14.

Conventional soybean prices rose to above $15 per bushel during 2011-14. Prices at this level would have severely limited organic price premiums had organic soybean prices not also increased sharply (fig. 6). The gap between average monthly conventional and organic soybean prices rose steadily from 2011 through 2012 with organic feed and food soybeans reaching about $30 per bushel, nearly $15 per bushel more than conventional soybean prices.

By the end of 2013, organic feed soybean prices were around $25 per bushel and food soybeans near $30 per bushel, creating price premiums for organic soybeans in the $11 to $16 per-bushel range. During 2014, conventional soybean prices fell to around $10 per bushel while organic prices remained in the $25-$30 per-bushel range, resulting in organic soybean price premiums of $15 to $20 per bushel during 2014.

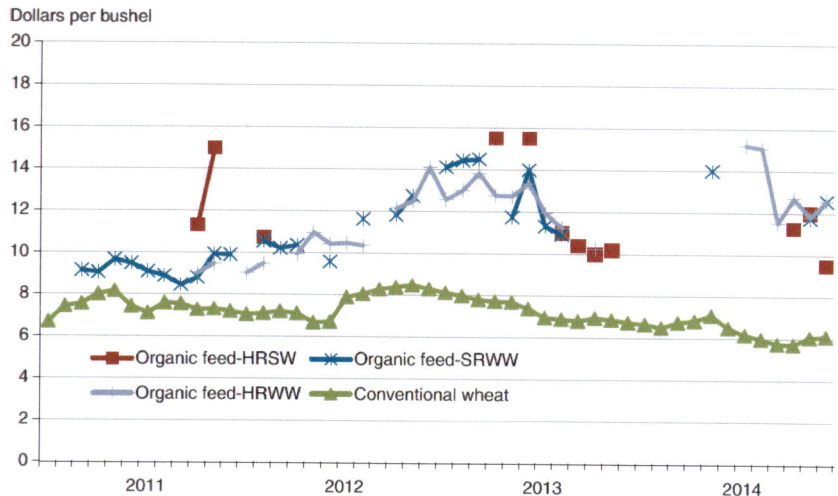

Note: HRSW = hard red spring wheat, SRWW = soft red winter wheat, HRWW = hard red winter wheat.

Source: Organic prices from USDA, Agricultural Marketing Service; conventional prices from USDA, National. Agricultural Statistics Service.

Figure 5. Organic feed and conventional wheat prices, 2011-14.

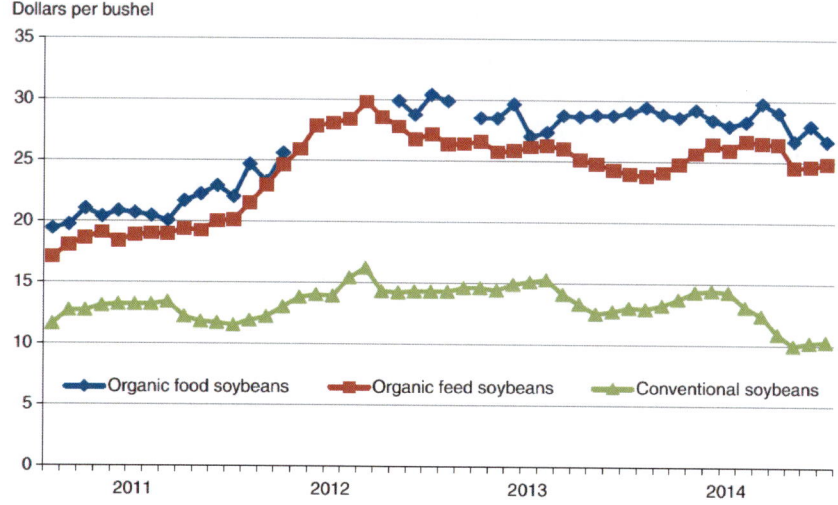

Source: Organic prices from USDA, Agricultural Marketing Service; conventional prices from USDA, National Agricultural Statistics Service.

Figure 6. Organic food and feed, and conventional soybean prices, 2011-14.

ORGANIC CROPPING SYSTEMS LITERATURE

Much of what is known about organic cropping systems stems from multidisciplinary research conducted with long-term experimental trials that compare the agronomic, economic, and sometimes environmental performance of organic and conventional systems. The identical weather and soil conditions under which field experiments are conducted provide opportunities not possible with observational studies, such as replication, precise field measurements, and long-term comparisons. In these types of studies, descriptive and analytical data are collected on crop yields and management practices, and the productivity, economic viability, and in some cases the potential environmental impacts of different farming systems are statistically assessed.

Previous research based on data from long-term experimental trails has shown mixed results when comparing the returns to organic production with those of conventional production, but the results have been generally favorable for organic systems. Several studies reported that organic production generated higher returns (Delate et al., 2003; Delate, 2013; Chavas et al., 2009; Clark, 2009) while others depended on whether historical organic premiums were to be paid (Mahoney, et al., 2001), were conditional on either the price premium and cropping system (Smith et al., 2004) or the size of farm (Delbridge et al., 2013), or depended on transition costs (Hansen et al., 1997). One study based on observational data from organic farms found significant variability in the production and financial performance of organic farms, much like that of conventional farms (Nordquist et al., 2014).

Prior studies also report mixed results concerning organic and conventional crop yields. Some of the experimental research indicates similar yields from conventional and organic systems (Delate et al., 2003; Delate et al., 2013; Pimentel et al., 2005), and potentially higher organic yields during drought years (Pimentel et al., 2005). Other studies have shown lower organic yields relative to conventional systems (Mahoney et al., 2001; Clark, 2009), but these lower yields were offset by lower production costs. Most studies have reported lower production costs from organic relative to conventional systems (Delate et al., 2003; Delate et al., 2013; Pimentel et al., 2005), although the cost analysis of organic systems has been primarily limited to a comparison of variable costs.

Long-term agricultural experiments have led to an improved understanding of the main biophysical and economic processes associated with different farming systems, addressing basic research questions about yields, profitability, and environmental impacts. In most of the situations studied, organic cropping systems generated returns above costs equal to or greater than those of conventional systems, sometimes generating much higher returns. Whether these results can be achieved outside of the experimental setting is uncertain mainly because organic production employs approaches to nutrient availability, pest control, and soil management that are profoundly different. These experiments also cannot account for the "human factor"—the valuable local knowledge and agricultural expertise that every farmer acquires through onfarm experience. The human factor plays a crucial role in organic farming.

Unlike most of the previous research, our study uses observational data where the treatment, choice of organic production or not, is not randomly assigned as in the experimental setting. Rather the observations "self-select" their status regarding the treatment. Crop producers themselves choose to produce certified organic crops rather than organic production being randomly assigned among producers. When assignment to the treatment is not random, simply comparing the effect on outcomes of the groups ignores underlying factors that influence both assignment to the treatment and the effect. For example, if crop producers' education level is correlated with both choice of organic certification and crop production costs, then the difference in crop production costs between the two groups may be due to both the treatment status and education level. Estimating the treatment-effect without controlling for potential covariate and sample-selection effects can lead to biased estimates (see "Appendix: Empirical Procedure").

Further, our study examines the relative profitability of organic and conventional field crop production defined as returns above various levels of production costs. Most prior studies define returns without economic costs for major resources including land, labor, and/or capital. This avoids critical assumptions about land rents, wage rates, and interest rates on borrowed capital. However, the usage and costs of these inputs can vary significantly between organic and conventional production systems. In this study, estimated returns above total economic costs, including charges for all resources used in production, are indicators of the relative profitability of organic field crop systems. This indicator provides information about the motivation for transitioning to organic production.

ORGANIC AND CONVENTIONAL FIELD CROP YIELDS

Organic and conventional crop yields reported in much of the published experimental research have been similar, but average organic yields in the ARMS data for each crop were significantly lower than those of conventional production. Unit production costs were computed as per-acre costs divided by the yield per acre of each crop. The average yield for organic corn was 118 bushels per acre in 2010, compared with 161 bushels for conventional corn. Organic wheat producers had an average yield of 30 bushels per acre in 2009, compared with 44 bushels for conventional production. Average yields for organic soybean producers in 2006 were also significantly lower, 31 versus 47 bushels per acre for conventional production. This amounts to an average yield penalty for organic production on commercial farms of 27 percent for corn, 32 percent for wheat, and 34 percent for soybeans.[3]

Previous research, based primarily on long-term cropping system data, suggests that significant returns are possible from organic crop production, often the result of obtaining similar conventional and organic yields with lower organic production costs. This study finds organic crop yields to be much lower than those of conventional production. The yield differences estimated from ARMS are similar to those estimated from the 2011 Organic Production Survey (USDA/NASS, c) relative to those from the 2011 Crop Production Report (USDA/NASS b). These 2011 data show organic corn yields to be 41 bushels per acre less than conventional yields, organic wheat yields to be 9 bushels per acre less than conventional yields, and organic soybean yields to be 12 bushels per acre less than conventional yields (fig. 7). The organic/conventional yield differences estimated from the ARMS data are slightly larger at 43, 14, and 16 bushels per acre, respectively, for each crop.[4]

As previously described, achieving yields was reported in the ARMS as one of the most difficult aspects of organic production. A reason for the yield differences measured with observational data may be the unique problems presented from implementing organic systems outside of the experimental setting, such as achieving effective weed control. Also, it is possible that the genetically modified conventional seed varieties that are commonly used for corn and soybean production are higher performing than standard organic seed varieties.

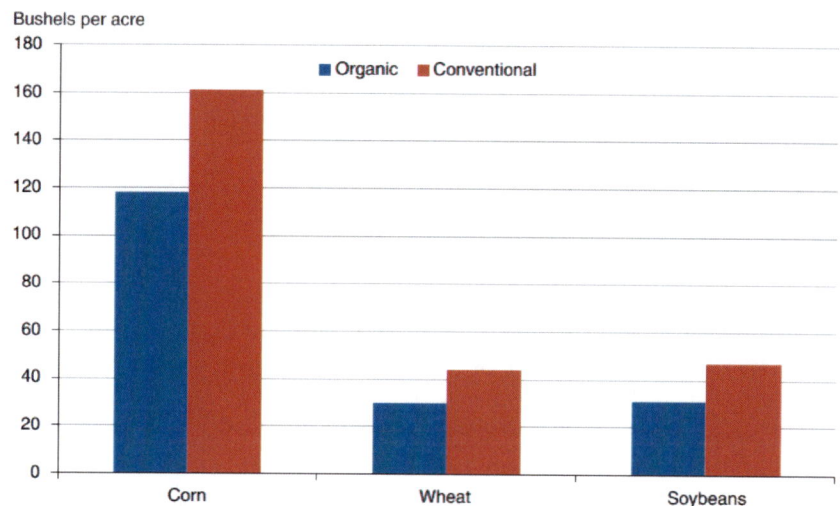

Source: USDA, National Agricultural Statistics Service, 2011 Certified Organic Production Survey and Crop Production: 2011 Summary.

Figure 7. Organic and conventional corn, wheat, and soybean yields, 2011.

CHARACTERISTICS AND COSTS OF ORGANIC AND CONVENTIONAL CROP FARMS

This section describes the similarities and differences between organic and conventional field crop producers using a statistical test of mean differences in farm, operator, and production c haracteristics for each group. Differences in mean operating, operating plus capital, and total economic costs per bushel are also statistically tested (see box, "Production Costs").

Corn

A summary of the 2010 ARMS corn producer data indicates that organic corn production was conducted on farms with less total acreage than conventional farms, and organic farms also harvested less corn acreage (table 1). Mean operator characteristics, including age and off-farm employment, were not statistically different between organic and conventional corn producers, but a lower percentage of organic producers had completed just high school (no college) than had conventional producers. Among regions, organic producers were more likely to be located in the Lake States (Michigan, Minnesota, and Wisconsin) and Northeast (New York and Pennsylvania) and less likely to be located in the Plains States (Kanasa, Nebraska, North Dakota, and South Dakota) than were conventional producers.

Production Costs

The average treatment-effect (ATE) of organic certification is measured using each technique (see "Appendix: Empirical Procedure") on three levels of unit (per bushel) production costs. Unit production costs are divided into operating costs, operating plus capital costs, and total economic costs. Operating costs include costs for seed; fertilizer; chemicals; custom operations; fuel, lubrication, and electricity; repairs; purchased irrigation water; hired labor; and operating interest. Capital costs include the annualized cost of maintaining the capital (economic depreciation and interest) used in production, estimated using the capital recovery approach, and costs for non-real estate property taxes and insurance. Total economic costs are the sum of operating and capital costs, plus opportunity costs for land and unpaid labor, and allocated costs for general farm overhead items. Costs of organic and conventional production are computed according to procedures used by USDA (USDA/ERS, 2012b).

Total operating costs is an indicator of the relative success of operations in terms of their ability to meet short-term financial obligations. The sum of operating and capital costs provides an indicator of whether operations can replace capital assets as needed and stay in business over time. Other costs are primarily opportunity costs of owned resources (land and labor) that may or may not influence production decisions. Opportunity costs of owned resources may vary significantly among producers and producers may be willing to accept returns to these resources different from assumed charges. Lifestyle preferences and costs of switching occupations, among other factors, affect producers' perceptions of their opportunity costs.

Table 1. Mean characteristics and practices of U.S. conventional and organic corn farms, 2010

Item	Type of farm	
	Organic (N=243)	Conventional (N=1,087)
Farm characteristics:		
Farm acres operated (per farm)	451	794**
Farm operator		
Off-farm occupation (percent)	11	18
Age (years)	51	56
Younger than 50 years old (percent)	44	30
Education (percent)		
Less than high school	24	8
Completed high school	29	45**
Attended college	47	47
Location (percent)		
Corn Belt (IL, IN, IA, MO, OH)	40	49
Lake States (MI, MN, WI)	40	24**
Northeast States (NY, PA)	14	6*
Plains States (KS, NE, ND, SD)	6	21**
Corn production practices:		
Harvested corn acres (per farm)	103	289**
Genetically modified seed (percent)	0	92**
Crop rotation (percent)		
Monoculture	0	0
Continuous row crop	17	77**
Idle year	35	10**
Other	48	13**
Field operations (percent)		
Moldboard plow	65	9**
No-till planter	5	35**
Row cultivator	68	5**
Other practices (percent)		
Irrigation	d	7**
Applied commercial fertilizer	51	97**
Applied manure or compost	75	22**
Corn yields, prices, and costs:		
Yield (bushels (bu) per acre)	118	161**
Price ($ per bu)	7.15	4.32**

Table 1. (Continued)

	Type of farm	
Item	Organic (N=243)	Conventional (N=1,087)
Operating costs ($ per bu)	1.75	1.80
Operating plus capital costs ($ per bu)	2.81	2.37**
Total economic cost ($ per bu)	4.68	3.43**

Note: Asterisks denote a statistically significant difference with the organic mean at the 10-percent (*) and 5-percent (**) levels.

d = insufficient data for legal disclosure.

Source: USDA, Economic Research Service and USDA, National Agricultural Statistics Service, 2010 Agricultural Resource Management Survey of corn producers.

Production practices used on organic corn farms varied significantly from those used on conventional corn farms. Most conventional producers planted corn in rotation with row crops, mainly soybeans.[5] Organic producers more often used an idle year and a meadow crop, such as alfalfa, in rotation with corn.[6] Organic corn producers used more intensive tillage practices, including moldboard plows, while conventional producers were more likely to use a no-till corn planter. Conventional producers relied heavily on various types of genetically modified seed and chemical weed control. Almost all conventional corn producers applied commercial fertilizers as opposed to organic producers who, instead, applied manure and/or compost.

Mean operating costs per acre for corn production were significantly less for organic (29 percent) than for conventional corn farms and operating plus capital costs were 13 percent less, while the difference in total economic costs per acre was not statistically significant. Conventional corn growers had significantly higher seed, fertilizer, and chemical costs than organic growers, but lower costs for fuel, repairs, capital, and labor as organic producers substituted manure and field operations for fertilizers and chemicals (fig. 8).[7,8] Paid and unpaid labor costs for organic production were significantly higher, totaling $71 per acre for organic corn versus $26 per acre for conventional corn. Total operating costs and operating plus capital costs per acre for organic corn were about $80 and $50 per acre lower, respectively, than for conventional corn.

Mean operating costs per bushel of corn were not significantly different between organic and conventional farms. However, operating plus capital costs and total economic costs per bushel were significantly higher among the organic corn farms (see table 1). Mean operating plus capital costs were more than 40 cents per bushel higher and mean total economic costs were $1.25 per bushel higher on organic compared with conventional corn farms. The average price reported as received for organic corn in 2010 was $7.15 per bushel, compared with a harvest-period price of $4.32 per bushel for conventional corn.[9] Both organic and conventional corn were profitable in 2010, but with an average organic price premium of $2.83 per bushel, mean returns above all costs were higher for organic than for conventional corn production.[10]

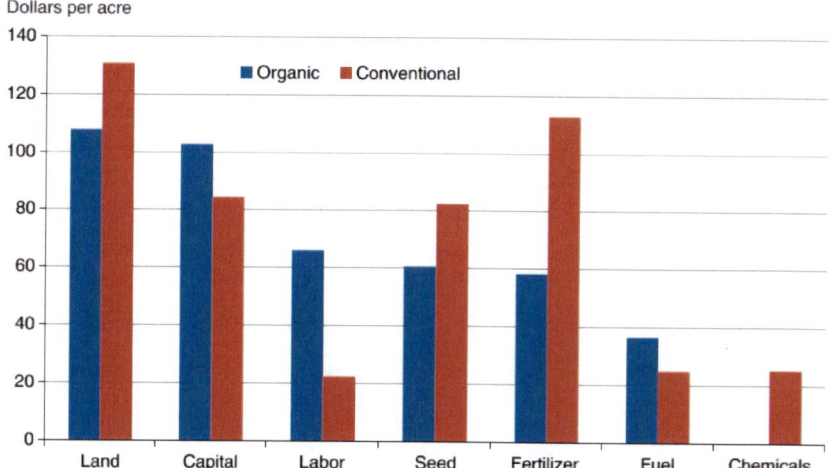

Note: Organic input costs are ordered from highest to lowest. Labor includes hired labor and unpaid labor costs. Source: USDA, Economic Research Service and USDA, National Agricultural Statistics Service, 2010 Agricultural Resource Management Survey.

Figure 8. Costs per acre of organic and conventional corn production by input, 2010.

Wheat

A summary of the 2009 ARMS data of wheat producers indicates that organic wheat production was conducted on farms similar in size to conventional producers, but organic farms harvested less wheat acreage (table 2). Organic wheat producers were younger, with a higher percentage younger than 50 years old, and a higher proportion of organic producers had attended college. Among wheat regions, organic farms were less likely to be located in the Southern Plains (Kansas, Oklahoma, and Texas) region, and were largely in the Northern Plains (Colorado, Montana, Nebraska, North Dakota, and South Dakota) region.[11]

Most conventional producers planted wheat in rotation with row and other small grain crops, particularly corn and soybeans, in contrast to organic producers, who used a more varied rotation of other crops. Organic producers more often tilled the soil with a moldboard plow, while conventional producers were more likely to sow wheat with a no-till planter. As with corn, most conventional producers applied commercial fertilizers to wheat, while organic producers applied manure or compost.

Average operating costs per acre for producing wheat were not significantly different between conventional and organic producers, but the composition of operating costs was very different (fig. 9). Conventional wheat growers had significantly higher fertilizer and chemical costs than organic growers, but lower costs for seed, fuel, and repairs as organic producers substituted manure and field operations for fertilizers and chemicals.[12] Total operating costs and operating plus capital costs per acre for organic wheat were about $20 per acre lower than for conventional wheat, but were not significantly different because of substantial variation in organic wheat production costs.[13] Average total economic costs of organic and conventional wheat production differed by only about $7 per acre.

Table 2. Mean characteristics and practices of U.S. conventional and organic wheat farms, 2009

	Type of farm	
Item	Organic (N=182)	Conventional (N=1,339)
Farm characteristics:		
Farm acres operated (per farm)	1,458	1,641
Farm operator		
Off-farm occupation (percent)	22	16
Age (years)	55	58
Younger than 50 years old (percent)	33	22*
Education (percent)		
Less than high school	3	4
Completed high school	24	37*
Attended college	73	60*
Location (percent)		
Central States (IL, MI, MN, MO, OH)	27	34
Northern Plains (CO, MT, NE, ND, SD)	42	34

	Type of farm	
Item	Organic (N=182)	Conventional (N=1,339)
Southern Plains (KS, OK, TX)	12	28**
Northwest (ID, OR, WA)	19	3
Wheat production practices:		
Harvested wheat acres (per farm)	258	405**
Purchased seed (percent)	50	55
Crop rotation (percent)		
Monoculture	1	3**
Continuous row crop/small grain	40	54*
Idle year	37	36
Other	22	8
Field operations (percent)		
Moldboard plow	30	5**
No-till planter	14	36**
Row cultivator	65	3**
Other practices (percent)		
Irrigation	21	3
Applied commercial fertilizer	17	84**
Applied manure or compost	37	6**
Wheat yields, prices, and costs:		
Yield (bushels (bu) per acre)	30	44**
Price ($ per bu)	9.30	5.51**
Operating costs ($ per bu)	3.10	2.65
Operating plus capital costs ($ per bu)	5.92	4.49**
Total economic cost ($ per bu)	9.11	6.07**

Note: Asterisks denote a statistically significant difference with the organic mean at the 10-percent (*) and 5-percent (**) levels. d = insufficient data for legal disclosure.

Source: USDA, Economic Research Service and USDA, National Agricultural Statistics Service, 2009 Agricultural Resource Management Survey of wheat producers.

With lower yields, average operating costs per bushel of organic wheat production were 45 cents higher, though not statistically different than those of conventional producers. Higher operating plus capital costs ($1.43 per bushel) and total economic costs ($3.04 per bushel) than for conventional wheat were statistically significant. The average price reported as received for organic wheat in 2009 was $9.30 per bushel, compared with a harvest-period price of $5.51 per bushel for conventional wheat, resulting in an average organic price premium of $3.79 per bushel.[14] Mean returns per bushel above all costs were positive for organic wheat but negative for conventional wheat production in 2009.

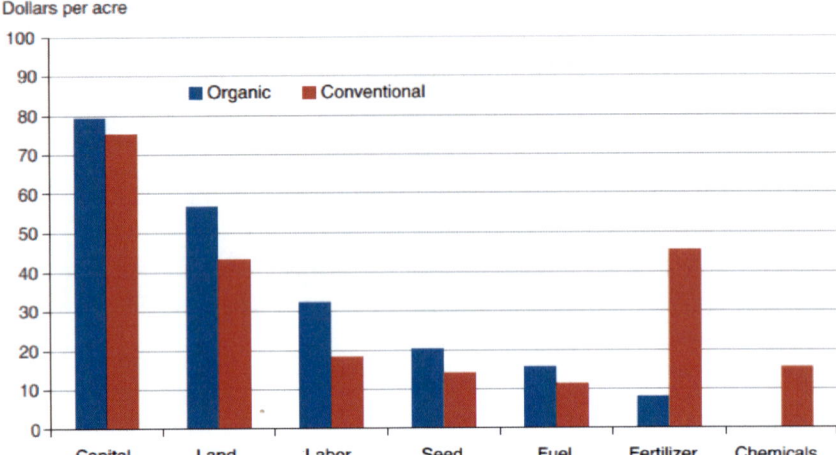

Note: Organic input costs are ordered from highest to lowest. Labor includes hired labor and unpaid labor costs. Source: USDA, Economic Research Service and USDA, National Agricultural Statistics Service, 2009 Agricultural Resource Management Survey.

Figure 9. Costs per acre of organic and conventional wheat production by input, 2009.

Soybeans

A summary of the 2006 ARMS data indicates that organic soybean production was conducted on smaller operations than conventional production and organic producers harvested fewer soybean acres (table 3). Despite their smaller size, organic soybean producers were less likely to report off-farm employment as their major occupation. The average age of organic and conventional producers was much the same with a similar distribution. Significantly more of the conventional producers reported just a high school education, while a higher percentage of organic producers had attended college. Organic producers were more often located in the Lake States and Plains States.

Nearly all conventional producers used genetically modified herbicide-tolerant seed, which is prohibited in organic production. Most organic producers used clear hilum seed. Clear hilum is a food-grade soybean most often used for soymilk and tofu production. Most organic producers planted soybeans in standard rows, while most conventional producers planted soybeans in narrow rows. This allowed organic producers to use a cultivator for weed control while conventional producers rarely used a cultivator.[15] Conventional producers mostly used a crop rotation comprised of continuous row crops, whereas organic producers more often rotated soybeans with small grain and meadow crops (e.g., alfalfa and other hay), and included an idle year in the rotation. Organic producers more often used intensive tillage practices, including moldboard plows. Conventional producers were much more likely to use a no-till planter.

Average soybean operating costs per acre were not significantly different among conventional and organic producers, but their composition was very different (fig. 10). Conventional production involved much higher chemical costs ($13.97 versus $0.02 per acre), while organic systems substituted field operations for chemicals and incurred much higher fuel, repairs, and labor costs. Capital costs were also much higher for organic production due to the greater use of field machinery. Paid and unpaid labor costs for organic production totaled $54 per acre, compared with $17 for conventional production. Total operating plus capital costs and total economic costs were significantly higher for organic production, averaging more than $30 and $60 per acre higher, respectively, than for conventional production.

With lower yields and higher per-acre costs, average operating costs per bushel for organic producers were $1.37 higher than for conventional producers, mean operating and capital costs were nearly $3 higher, and mean total economic costs were more than $5 higher (table 3).[16] The average price premium received by organic producers was more than $9 per bushel in 2006, making organic soybeans profitable on average, while returns to conventional soybean production were negative.

Table 3. Mean characteristics and practices of U.S. conventional and organic soybean farms, 2006

	Type of farm	
Item	Organic ($N=237$)	Conventional ($N=1,425$)
Farm characteristics:		
Farm acres operated (per farm)	478	748**
Farm operator		
Off-farm occupation (percent)	16	26**

Table 3. (Continued)

Item	Type of farm	
	Organic (N=237)	Conventional (N=1,425)
Age (years)	54	55
Younger than 50 years old (percent)	32	32
Education (percent)		
Less than high school	18	5*
Completed high school	24	46**
Attended college	51	24**
Location (percent)		
Corn Belt (IL, IN, IA, MO, OH)	42	56
Lake States (MI, MN, WI)	51	24**
Plains States (KS, NE, ND, SD)	20	7**
Soybean production practices:		
Harvested soybean acres (per farm)	117	272**
Genetically modified seed (percent)	0	97**
Crop rotation (percent)		
Monoculture	1	4**
Continuous row crop	19	79**
Idle year	40	9**
Other	17	4**
Field operations (percent)		
Moldboard plow	36	5**
No-till planter	6	50**
Row cultivator	65	3**
Other practices (percent)		
Irrigation	3	5
Applied commercial fertilizer	7	32**
Applied manure or compost	28	7**
Soybean yields, prices, and costs:		
Yield (bushels (bu) per acre)	31	47**
Price ($ per bu)	14.64	5.48**
Operating costs ($ per bu)	3.32	1.95**
Operating plus capital costs ($ per bu)	6.24	3.40**
Total economic cost ($ per bu)	10.97	5.87**

Note: Asterisks denote a statistically significant difference with the organic mean at the 10-percent (*) and 5-percent (**) levels. d = insufficient data for legal disclosure.

Source: USDA, Economic Research Service and USDA, National Agricultural Statistics Service, 2006 Agricultural Resource Management Survey of soybean producers.

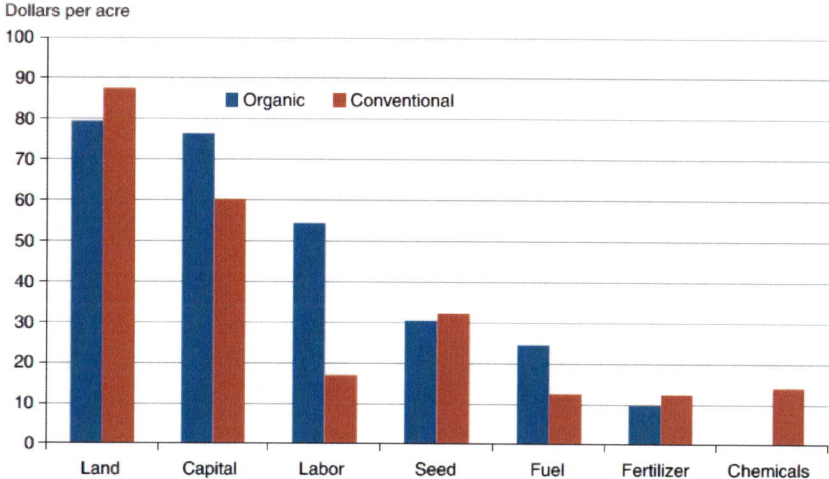

Note: Organic input costs are ordered from highest to lowest. Labor includes hired labor and unpaid labor costs. Source: USDA, Economic Research Service and USDA, National Agricultural Statistics Service, 2006 Agricultural Resource Management Survey.

Figure 10. Costs per acre of organic and conventional soybean production by input, 2006.

ORGANIC AND CONVENTIONAL PRODUCTION COST DIFFERENCES

The primary innovation presented in the empirical section of this study is the development of two treatment-effect estimators: propensity-score matching and linear regression with endogenous treatment-effects (see "Appendix: Empirical Procedure"). Both estimators provide measures of the production cost differences between organic and conventional field crops. The results are compared with the mean difference shown in the previous section for each crop.

The method of propensity-score matching involved the estimation of a binary choice model to compute propensity scores, indicating the likelihood of a producer being organic. Propensity-scores were used to match similar conventional with organic producers of each crop. The dependent variable in the model was treatment status, certified organic or not. The set of independent variables in the model were farm and operator characteristics and commodity production practices (see appendix tables 1, 3, and 5), selected based on findings in the review of literature and information about differences between organic and conventional farms gleaned from the analysis of means. For example, crop acreage, operator education, and crop rotation variables were used in the models to determine the propensity-scores. After computing the propensity-score for all observations, each organic producer was matched with the conventional producer having the most similar propensityscore.[17] Differences in the production costs across the sample of matched organic and conventional producers determined the treatment-effects.

The model of linear regression estimates with endogenous treatment-effects was specified differently. The linear regression model was comprised of two equations, a treatment (or participation) equation and an effect (or outcome) equation estimated simultaneously. The treatment equation was specified with exogenous variables expected to influence choice of the treatment (certified organic or not), in contrast to the propensity-score model that was specified to find the best match of farms based on observable variables.[18] For example, exogenous farm and operator variables were included in the treatment equations while variables endogenous to the treatment, such as production practice variables, were included in the effect equations (see appendix tables 2, 4, and 6).[19] Parameter estimates from the model equations were used to compute treatment-effects resulting from the regression approach.

Average treatment-effects on each of the cost-of-production variables from the two estimators for each crop are shown in table 4. The mean difference in the production costs of organic and conventional producers is shown for comparison. Among corn growers, both treatment-effect models produced estimates of the impact of organic certification on production costs that were much greater than the mean difference. Despite using very different approaches, the estimated difference for operating costs and for operating plus capital costs were similar for each treatment-effect model, although those estimated with the propensity-score matching estimator were not statistically significant.[20] These estimated treatment-effects differed by less than 5 cents per bushel.

Table 4. U.S. corn, wheat, and soybean crops: Organic compared with conventional production costs per bushel using alternative estimators[1]

Crop/cost item	Mean difference	Estimator Propensity-score matching	Regression w/endogenous treatment-effects
		$ per bushel	
Corn			
Operating costs	-0.05	0.26	0.22**
Operating plus capital costs	0.44**	0.92	0.94**
Total economic costs	1.25**	1.62**	1.93**
Wheat			
Operating costs	0.45	0.37	0.83
Operating plus capital costs	1.43**	1.82	2.00**
Total economic costs	3.04**	3.38	3.87**
Soybeans			
Operating costs	1.37**	0.48	1.05**
Operating plus capital costs	2.84**	2.26**	2.78**
Total economic costs	5.10**	5.53**	6.59**

[1] Mean difference, propensity-score matching, and regression with endogenous treatment-effect estimators.

Note: Estimates show the difference in costs of production between certified organic producers and conventional producers using each type of estimator. * and ** denote statistical significance at the 10-percent and 5-percent levels, respectively, for each estimator.

Source: USDA, Economic Research Service and USDA, National Agricultural Statistics Service, Agricultural Resource Management Survey: 2010 for corn, 2009 for wheat, and 2006 for soybeans.

The estimated average treatment-effect on total economic costs for corn was statistically significant with both estimators, at a value of $1.62 per bushel using the propensity-score estimator and $1.93 per bushel using the regression with endogenous treatment-effect estimator. Both of these estimates were greater than the mean difference in total economic costs of corn production, suggesting that not adjusting for the influence of covariates would have contributed to an underestimation of the treatment-effect. The propensity-score matching model estimates that total economic costs among organic corn growers are 30 percent higher than that reflected by the mean difference. Total economic costs are estimated to be 54 percent higher with the regression model.

Average treatment-effects on each of the cost-of-production variables from the two treatment-effect estimators for wheat are shown in the middle of table 4. None of the differences in wheat operating costs were statistically significant. Like the findings for corn, both treatment-effect models produced estimates of the impact of organic certification on operating plus capital and total economic costs of wheat production that were larger than the mean difference. The estimated average treatment-effect on total economic costs was statistically significant with the regression with endogenous treatment-effects estimator, at a value of $3.87 per bushel, more than 25 percent higher than the mean difference. The propensity-score estimator measured the treatment-effect at $3.38 per bushel, also greater than the mean difference, but not statistically significant.

The estimated average treatment-effect on each cost-of-production variable for soybeans is shown in the bottom part of table 4. Nearly all estimates of the difference between organic and conventional soybean production costs were statistically significant. Unlike the estimators for other crops, the mean difference in operating and operating plus capital costs between organic and conventional soybean producers was greater than that from either treatment-effect estimator. Both treatment-effect estimators for total economic costs were greater than the mean difference, again suggesting that the mean difference underestimates the treatment-effect. The propensity-score matching model estimates that total economic costs among soybean growers are 8 percent higher than the mean difference. Total economic costs are estimated at nearly 30 percent higher with the regression model.

ORGANIC TRANSITION AND CERTIFICATION COSTS

The estimated cost differences indicate the additional costs incurred by operations producing organic relative to conventional production, but do not include the costs associated with the transition to organic production or the costs associated with annual third-party certification fees. Before an operation is certified to sell organic crops, the cropland must be managed organically for a minimum of 36 months (USDA/AMS a). This means that operations must undergo 2 years of organic production costs before selling crops as certified organic at the end of the third year of transition.[21] In order to maintain the organic certification, the producer must pay annual certification costs, including costs for items such as application fees, renewal fees, assessment of annual production or sales, and inspection fees.

Higher costs for 2 years can be considered as an investment necessary to return higher organic prices over the planning horizon of the organic operation. The investment was determined by the estimated additional costs incurred by organic operations as indicated by the mean difference and from each treatment-effect model estimator for 2 years of the 3-year transition period in which organic price premiums cannot be obtained. The annualized cost of this investment was computed using the capital recovery approach like the other capital costs. The investment was spread over an assumed planning horizon for organic production of 20 years.[22] The addition of estimated transition and certification costs to the additional costs of producing organic compared with conventional corn, wheat, and soybean production using the mean difference and the two treatment-effect estimators are shown in table 5.[23]

Annual third-party certification costs were charged at the survey mean for each crop, adding to operating costs $0.10 per bushel for corn, $0.12 per bushel for wheat, and $0.41 per bushel for soybeans. Transition costs added between $0.08 and $0.17 per bushel to operating plus capital costs, and $0.23 to $0.35 per bushel to total economic costs among the various estimators for corn production. Total economic costs were estimated at $1.92 per bushel higher for certified organic corn compared with conventional corn production using the propensity-score matching estimator, and $2.27 per bushel higher using the regression with endogenous treatment-effects estimator, both much higher than the mean difference of $1.50 per bushel.

Transition costs added between $0.25 and $0.45 per bushel to operating plus capital costs, and $0.46 to $0.82 per bushel to total economic costs among the various estimators for wheat production. Like corn, both treatment-effect estimators indicated much higher additional costs associated with organic production ($3.90 and $4.46 per bushel) than did the mean difference of $3.53 per bushel. For soybeans, transition costs added between $0.43 and $0.52 per bushel to operating plus capital costs, and $0.94 to $1.21 per bushel to total economic costs among the estimators. Total economic costs for soybeans with each estimator showed the same pattern as with corn and wheat, with both treatment-effect estimators indicating a greater difference between organic and conventional soybean costs than the mean difference. Total economic costs were estimated at $6.62 per bushel higher for certified organic soybean production using the propensity-score matching estimator, and $7.81 per bushel higher using the regression with endogenous treatment-effects estimator, both higher than the $6.13-per-bushel increase suggested by the mean difference.

Table 5. U.S. corn, wheat, and soybean crops: Organic compared with conventional production costs per bushel, including organic transition and certification costs, by estimator[1]

Crop/cost item	Mean difference	Estimator	
		Propensity-score matching	Regression w/ endogenous treatment-effects
		$ per bushel	
Corn			
Operating costs	0.05	0.36	0.32
Operating plus capital costs	0.59	1.13	1.15
Total economic costs	1.50	1.92	2.27
Wheat			
Operating costs	0.57	0.49	0.95
Operating plus capital costs	1.72	2.16	2.36
Total economic costs	3.53	3.90	4.46
Soybeans			
Operating costs	1.78	0.89	1.46
Operating plus capital costs	3.60	2.95	3.53
Total economic costs	6.13	6.62	7.81

[1]Estimates show the difference in costs of production between certified organic producers and conventional producers using each type of estimator including the addition of transaction and certification cost estimates. Transition costs are treated as a capital investment necessary to return the higher organic crop price over the planning horizon of the operation, and thus are not part of annual operating costs. Certification costs are an annual operating expense charged at the mean estimate of 10 cents per bushel for corn, 12 cents per bushel for wheat, and 41 cents per bushel for soybeans.

Source: USDA, Economic Research Service and USDA, National Agricultural Statistics Service, Agricultural Resource Management Survey: 2010 for corn, 2009 for wheat, and 2006 for soybeans.

A comparison of the percentage increase in total economic costs of organic production, from each estimator, relative to the mean total economic costs of conventional production for corn, wheat, and soybeans is shown in figure 11. On a percentage basis, being organic increases total economic costs the most for soybeans where organic production costs are more than double conventional production costs regardless of estimator. Organic production raises the cost of corn and wheat production by roughly 50 to 75 percent depending upon the estimator. Also, both treatment-effect estimators resulted in total economic costs that were higher than the mean difference among all the crops, both running between about 10 and 30 percent higher than the mean with the largest difference from the regression with endogenous treatment-effects estimator.

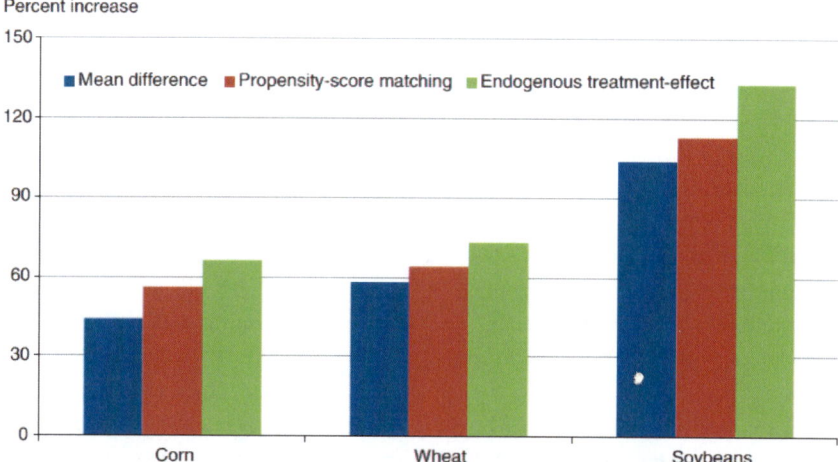

Note: Percentage increase in the total economic cost of organic production by estimator relative to the mean total economic cost of conventional production.
Source: USDA, Economic Research Service and USDA, National Agricultural Statistics Service, 2006, 2009, and 2010 Agricultural Resource Management Survey.

Figure 11. Increase in total economic costs including transition and certification costs: Organic versus conventional field crop production by estimator.

RETURNS TO ORGANIC FIELD CROP PRODUCTION

Comparison of the additional costs associated with organic production with historic price premiums provides an indication of the returns associated with organic field crop production. Average organic corn prices were $7.15

per bushel among the producers sampled in 2010, compared with $4.32 per bushel for conventional producers, a $2.83-per-bushel price premium. This premium is greater than both treatment-effect estimates of higher corn costs plus organic transition and certification costs ($1.92 and $2.27 per bushel), suggesting an average margin of $0.56 to more than $0.90 per bushel above total economic costs in 2010 for organic corn production.

Higher conventional corn prices that rose to between $6 and $8 per bushel during 2011-13 would have reduced this margin had organic corn prices not also increased (see figure 3). The gap between average organic and conventional corn prices rose steadily from 2011 through 2012 with organic feed and food corn reaching above $16 per bushel, about $9-$10 higher than conventional corn prices. Organic corn prices declined in 2013, but the price differential between organic and conventional corn remained in the $5- to $7-per-bushel range, still much higher than the differential suggested by the cost analysis. Organic corn prices rebounded in 2014, while conventional corn prices moved lower, and the organic price premium returned to the $9-$10 per bushel range.

Higher organic wheat production costs compare to an average price premium of $3.79 per bushel for organic wheat in 2009, indicating tighter margins for organic wheat production than for organic corn. The 2009 price premium was sufficient to cover the additional operating plus capital costs associated with organic wheat, but did not cover the higher average total economic costs indicated by the cost analysis using both treatment-effect estimators ($3.90 and $4.46 per bushel). The gap between average organic and conventional wheat prices depended on the type of wheat produced. Between 2011 and 2013, price premiums for organic food wheat were generally higher than for organic feed wheat by about $2-$6 per bushel, and grew even wider in 2014 (see figures 4 and 5). Price premiums for organic food wheat were generally above the estimated total economic cost differential between organic and conventional wheat production. Farm prices received for organic feed wheat varied significantly during 2011-14, but much of the time were only $1-$4 per bushel higher than those for conventional wheat. This price differential was often below the additional economic costs of organic wheat production.

The additional costs for organic soybean production compares to an average price premium of $9.16 per bushel for organic soybeans in 2006. This suggests that organic soybean producers, on average, earned returns above operating plus capital costs in the range of about $5.60 to $6.20 per bushel, and returns above total economic costs ranging from about $1.35 to $2.50 per bushel in 2006, based on the treatment-effect estimators. Conventional soybean prices rose to above $15 per bushel during 2011-13. Prices at this level would have greatly reduced any organic price premium had organic soybean prices not also increased (see figure 6).

The gap between average organic and conventional soybean prices rose steadily from 2011 through 2012. Organic food and feed soybean prices both reached about $30 per bushel, nearly $15 per bushel more than conventional soybean prices. By the end of 2013, organic feed soybean prices were around $25 per bushel and food soybeans near $30 per bushel, creating price premiums high enough in both cases to easily cover the estimated additional costs of organic soybean production. Conventional soybean prices declined in 2014, while organic prices were mostly stable, resulting in even higher price premiums to organic soybean production in 2014.

The yield, price, and cost differences estimated in this study were used to compute the per-acre returns to organic versus conventional production for the survey year of each crop. Average additional costs of about $83-$98 per acre for corn, $55-$62 per acre for wheat, and $106-$125 for soybeans would be incurred from organic production (table 6). These cost estimates are based on the yield and cost differences estimated from the survey and include an annualized share of organic transition costs plus annual organic certification costs. Estimates of the difference in net returns per acre for organic versus conventional production are positive for corn ($66 and $51 per acre) and soybeans ($41 and $22), but negative for wheat (-$2 and -$9 per acre) using estimates from each of the treatment-effect estimators.

Table 6. U.S. corn, wheat, and soybean crops: Organic compared with conventional economic costs and returns, by estimator[1]

Crop/cost item		Estimator	
	Mean difference	Propensity-score matching	Regression w/ endogenous treatment-effects
		$ per acre	
Corn			
Gross value of production	148.18	148.18	148.18
Total economic costs	64.50	82.56	97.61
Net value of production	83.68	65.62	50.57
Wheat			
Gross value of production	53.06	53.06	53.06
Total economic costs	49.42	54.60	62.44

Table 6. (Continued)

Crop/cost item	Mean difference	Estimator Propensity-score matching	Regression w/ endogenous treatment-effects
Net value of production	3.64	-1.54	-9.38
Soybeans			
Gross value of production	146.56	146.56	146.56
Total economic costs	98.08	105.92	124.96
Net value of production	48.48	40.64	21.60

[1] Estimates show the difference in costs (including organic transaction and certification cost estimates) and returns of crop production between certified organic producers and conventional producers using each type of estimator. The difference in gross value of production was computed using the difference in mean yield and prices received for each crop. The difference in total economic costs is that estimated with each estimator times the difference in mean yield per acre for each crop.

Source: USDA, Economic Research Service and USDA, National Agricultural Statistics Service, Agricultural Resource Management Survey: 2010 for corn, 2009 for wheat, and 2006 for soybeans.

CONCLUSION

This study takes advantage of unique and detailed data collected in economic surveys of U.S. corn, wheat, and soybean producers. The data were unique because a targeted survey of organic producers sampled at a much higher rate than their occurrence in the population was included along with data from conventional producers. This allowed for an observational analysis of cost-of-production differences between conventional and organic crop production systems. These differences provide information about price premiums needed to make organic systems competitive with conventional systems, and about the additional costs incurred by producers transitioning to organic production.

Comparison of the treatment-effect estimators with mean differences suggests that estimates of the cost difference between organic and conventional production based on the mean likely understate actual cost differences. Higher average economic cost differences between organic and conventional corn, wheat, and soybean producers were estimated using two different treatment-effect models than those revealed from a comparison of means. Differences in observable covariates specified in the treatment-effect models account for some of the difference. The difference in costs may also be partly due to sample selection-bias (see "Appendix: Empirical Procedure"), addressed in the treatment-effect models.[24]

The main reason that organic returns were higher than conventional returns in the analysis of the ARMS data was not higher organic yields or lower per-acre production costs, but rather the price premiums paid for organic crops. Average organic corn and soybean prices were more than enough among the sampled producers to cover the higher total economic costs of organic production, including an annual prorated share of transition costs and annual certification costs. Organic food wheat prices were also sufficient to cover the higher costs of organic wheat production, while organic feed wheat prices were high enough to cover the additional operating plus capital costs, but not high enough to cover the additional total economic costs.

Price premiums received for organic crops were generally above the estimated additional costs of organic production for most crops during much of 2011-14. Estimates of the difference in net returns per acre for organic versus conventional production showed positive economic profit for corn and soybeans, consistent with expanded, or stable, organic acreage of each crop in recent years. Estimates of economic loss per acre, on average, for organic versus conventional wheat, mainly feed wheat, are consistent with the decline in organic wheat acreage shown in figure 1.

An implication of these results is that conventional farms may be able to earn greater returns above economic costs if transitioned to organic production. Despite potentially higher returns from organic production, the adoption of the organic approach among U.S. field crop producers remains extremely low. One possible reason there is not more organic production is the ease of producing for the conventional market. Seed and chemicals are readily available from local seed and chemical company dealers, and from markets at the local elevator. Organic farmers, in contrast, have to secure organic seed; learn to manage soil fertility, weeds, and other pests through natural methods; and find their own markets, which may require storage on the farm until pickup. Thus organic farming requires more onfarm management.

The low level of U.S. organic-crop adoption may also be due to variations in climatic and market conditions. Organic production is more attractive where crop pests are fewer, such as in northern States. Also, a market for the more expensive organic food or feed crops is needed, such as the demand for organic feed ingredients, both grain and silage, from the significant organic dairy industry that has developed in States in the upper Midwest and

Northeast.[25] Kuminoff and Wossink (2010) point out that sunk organic production costs associated with transition, which cannot be recovered if organic prices drop substantially, coupled with uncertainty about future returns, may help explain why there is so little organic acreage for major field crops in the United States.

Results of this study need to be considered in light of the limitations of the data and methods used. Only 1 year of production cost data were available for each crop. The production of organic crops, in contrast to conventional crops, is often part of a multiyear rotation of crop enterprises and idled land. Also, farm practice data were analyzed from a nonrandomized setting. Despite efforts to deal with these limitations using the treatment-effect models, the results may still be affected by residual confounding in which factors influence choice of the treatment and the effect in question. This potential confounding bias limits the causal inference and renders this primarily a study of association.

Further research could improve upon this study by evaluating onfarm organic production in a multiyear systems setting. Organic field crop producers may rotate with less profitable enterprises lowering overall cropping system returns, or the synergism associated with the management of multiple crop enterprises may result in greater returns than indicated by this single-year, single-enterprise analysis. A more thorough study of the economic returns to organic systems would account for the inherent multiyear nature of organic cropping systems.

REFERENCES

Abadie, A., and G.W. Imbens. 2012. *Matching on the Estimated Propensity Score*, Harvard University and National Bureau of Economic Research. http://www.hks.harvard.edu/fs/aabadie/ pscore.pdf

American Agricultural Economics Association. 2000. *Commodity Costs and Returns Estimation Handbook: A Report of the AAEA Task Force on Commodity Costs and Returns*. Ames, Iowa. ftp://ftp-fc.sc.egov.usda.gov/Economics/care/AAEA/PDF/AAEA%20Handbook.pdf

Becker, S.O., and M. Caliendo. 2007. "Sensitivity Analysis for Average Treatment Effects," *Stata Journal* 7(2007):71-83.

Becker, S.O., and A. Ichino. 2002. "Estimation of Average Treatment Effects Based on Propensity Scores," *Stata Journal* 2(2002):358-77.

Caliendo, M., and S. Kopeing. 2008. "Some Practical Guidance for the Implementation of Propensity Score Matching," *Journal of Economic Surveys* 22(2008):31-72.

Cavigelli, M., S. Mirsky, J. Teasdale, J. Spargo, and J. Doran. 2013. "Organic Grain Cropping Systems to Enhance Ecosystem Services," *Renewable Agriculture and Food Systems* 28(2013):145-159.

Chavas, J.P., J.L. Posner, and J.L. Hedtcke. 2009. "Organic and Conventional Production Systems in the Wisconsin Integrated Cropping Systems Trial: II. Economic and Risk Analysis 1993-2006." *Agronomy Journal* 101(2009):288-295.

Clark, S.F. 2009. "The Profitability of Transitioning to Organic Grain Crops in Indiana," *American Journal of Agricultural Economics* 91(2009)5:1497-1504.

Coulter, J., T. Delbridge, R. King, D. Allan, and C. Sheaffer. 2013. "Productivity, Economics, and Soil Quality in the Minnesota Variable-Input Cropping Systems Trial," *Crop Management* 12(2013).

Delate, K., M. Duffy, C. Chase, A. Holste, H. Friedrich, and N. Wantate. 2003. "An Economic Comparison of Organic and Conventional Grain Crops in a Long-Term Agroecological Research (LTAR) Site in Iowa," *The American Journal of Alternative Agriculture* 18(2003):59-69.

Delate, K., C. Chase, M. Duffy, C. Chase, and R. Turnbull. 2006. "Transitioning into Organic Grain Production: An Economic Perspective," *Crop Management* 5(2006).

Delate, K., C. Cambardella, C. Chase, A. Johanns, and R. Turnbull. 2013. "The Long-Term Agroecological Research (LTAR) Experiment Supports Organic Yields, Soil Quality, and Economic Performance in Iowa," proceedings from the USDA Organic Farming Systems Research Conference published in *Crop Management* (special issue 2013).

Delbridge, T.A., C. Fernholz, R. King, and W. Lazarus. 2013. "A Whole-Farm Profitability Analysis of Organic and Conventional Cropping Systems," *Agricultural Systems* 122(2013):1-10.

Federal Register. December 21, 2000. National Organic Program. FR 65(246): 80548-80684. http:// www.federalregister.com/Browse/Document/usa/na/ fr/2000/12/21/00-32257

Fitzgerald, J., P. Gottschalk, and R. Moffit. 1998. "An Analysis of Sample Attrition in Panel Data: The Michigan Panel Study of Income Dynamics," *The Journal of Human Resources* 33(1998):251-99.

Greene, C. and A. Kremen. 2003. *U.S. Organic Farming in 2000-2001*, U.S. Department of Agriculture, Economic Research Service, AIB-780.

Greene, W. H. 2000. *Econometric Analysis*, 4th edition. New Jersey: Prentice Hall.

Gunsolus, J.L. 2013. *Herbicide Resistant Weeds*, University of Minnesota Extension. http://www. extension.umn.edu/agriculture/crops/weed-management/herbicide-resistant-weeds

Hanson, J.C., E. Lichtenberg, and S.E. Peters. 1997. "Organic Versus Conventional Grain Production in the Mid-Atlantic: An Economic and Farming System Overview," *The American Journal of Alternative Agriculture* 12(1997):2-9.

Heckman, J., and S. Navarro-Lozano. 2004. "Using Matching Instrumental Variables, and Control Functions to Estimate Economic Choice Models," *Review of Economics and Statistics* 86(2004):30-57.

Heckman, J., H. Ichimura, and P. Todd. 1998. "Matching as an Econometric Evaluation Estimator," *The Review of Economic Studies* 65(1998):262-94.

Heckman, J.J. 1976. "The Common Structure of Statistical Models of Truncation, Sample Selection, and Limited Dependent Variables and a Simple Estimator for Such Models," *Annals of Economic and Social Measurement* 5(1976):475-91.

Imbens, G.W., and J.M. Wooldridge. 2009. "Recent Developments in the Econometrics of Program Evaluation," *Journal of Economic Literature* 47(2009):5-86.

Imbens, G.W. "Nonparametric Estimation of Average Treatment Effects Under Exogeneity: A Review." *Review of Economics and Statistics* 86(2004):4-29.

Kuminoff, N.V., and A. Wossink. 2010. "Why Isn't More U.S. Farmland Organic?," *Journal of Agricultural Economics* 61(2010):240-258.

Liu, X., and L. Lynch. 2007. *Do Agricultural Preservation Programs Affect Farmland Conversion*, selected paper prepared for presentation at the Agricultural and Applied Economics Association Annual Meeting, Portland, OR (July 29-August 1, 2007).

Mahoney, P.R., K.D. Olson, P.M. Porter, D.R. Huggins, C. A. Perillo, and R. K. Crookston. 2001. "Profitability of Organic Cropping Systems in Southwestern Minnesota," *Renewable Agriculture and Food Systems* 19(2001):35-46.

Mayen, C.D., J.V. Balagtas, and C.E. Alexander. 2010. "Technology Adoption and Technical Efficiency: Organic and Conventional Dairy Farms in the United States," *American Journal of Agricultural Economics* 92(2010):181-195.

McBride, W.D., and C. Greene. 2009. "The Profitability of Organic Soybean Production," *Renewable Agriculture and Food Systems* 24(2009):276-284.

Morgan S.L., and D.J. Harding. 2006. "Matching Estimators of Casual Effects," *Sociological Methods & Research* 35(2006):3-60.

Nannicini, T. 2001. "Simulation–Based Sensitivity Analysis for Matching Estimators," *Stata Journal* 7(2001):334-50.

Nichols, A. 2007. "Causal Inference with Observational Data," *Stata Journal* 7(2007):507-41.

Nordquist, D., R. Dvergsten, R. Joerger, and M. Moynihan. 2012. *2011 Organic Farm Performance in Minnesota*. Minnesota Department of Agriculture. http://www.mda.state.mn.us/~/media/Files/food/organic growing/orgfarmperf2011.ashx

Pimentel, D., P. Hepperly, J. Hanson, D. Doubs, and R. Seidel. 2005. "Environmental, Energetic, and Economic Comparisons of Organic and Conventional Farming Systems," *BioScience* 55(2005):573-582.

Pufahl, A., and C.R. Weiss. 2009. "Evaluating the Effects of Farm Programmes: Results From Propensity Score Matching," *European Review of Agricultural Economics* 36(2009):79-101.

Rosenbaum, P.R., and D.B. Rubin. 1983. "The Central Role of the Propensity Score in Observational Studies for Causal Effects," *Biometrika* 70(1983):41-55.

Smith, E.G., J.M. Clapperton, and R.E. Blackshaw. 2004. "Profitability and Risk of Organic Production Systems in the Northern Great Plains," *Renewable Agriculture and Food Systems* 19(2004):152-58.

Taur, L.W. 2009. "Estimation of Treatment Effects of Recombinant Bovine Somatotropin Using Matching Samples," *Applied Economic Perspectives and Policy* 31(2009):411-23.

Uematsu, H., and A.K. Mishra. 2012. "Organic Farmers or Conventional Farmers: Where's the Money?" *Ecological Economics* 78(2012):55-62.

U.S. Department of Agriculture, Agricultural Marketing Service. Various years. (USDA/AMS a). *National Organic Program*. http://www.ams.usda.gov/AMSv1.0/nop

U.S. Department of Agriculture, Agricultural Marketing Service. Various years. (USDA/ AMS b). *Organic Certification Cost Share Programs*. http://www.ams.usda.gov/AMSv1.0/ NOPCostSharing

U.S. Department of Agriculture, Agricultural Marketing Service. Various years. (USDA/AMS c). *Livestock & Grain Market News*. http://marketnews.usda.gov/portal/lg

U.S. Department of Agriculture, Economic Research Service. Various years. (USDA/ERS a). *Organic Production.* http://www.ers.usda.gov/Data/Organic

U.S. Department of Agriculture, Economic Research Service. Various years (USDA/ERS b). *Commodity Costs and Returns.* http://www.ers.usda.gov/data-products/commodity-costs-andreturns.aspx

U.S. Department of Agriculture, National Agricultural Statistics Service. Various issues. (USDA/ NASS a). *Agricultural Prices.* http://www.nass.usda.gov/Surveys/Guide_to_NASS_Surveys/
Prices_Received_and_Prices_Received_Indexes/index.asp

U.S. Department of Agriculture, National Agricultural Statistics Service. Various issues. (USDA/ NASS b). *Crop Production Annual Summary.* http://usda.mannlib.cornell.edu/MannUsda/view-DocumentInfo.do;jsessionid=C3A3E0254A2D10FC156C21493D06AB9E?documentID=1047

U.S. Department of Agriculture, National Agricultural Statistics Service. Various issues. (USDA/ NASS c) *2011 Certified Organic Production Survey.* http://usda.mannlib.cornell.edu/MannUsda/ viewDocumentInfo.do?documentID=1859

U.S. Department of Agriculture, Natural Resources Conservation Service. Various years. EQIP Organic Initiative. http://www.nrcs.usda.gov/wps/portal/nrcs/detail/national/ programs/?cid=nrcs143_008224

Wooldridge, J.M. 2010. *Econometric Analysis of Cross Section and Panel Data*, 2nd edition. The MIT Press. 2010.

APPENDIX: EMPIRICAL PROCEDURE

An objective of this study is to estimate the average treatment-effect (ATE) of organic certification on crop production costs. Estimation of the treatment-effect under nonexperimental settings, using observational data, is popular in social science research where experimental settings are not usually possible. A number of studies have provided the theoretical background (Heckman et al., 1998; Imbens, 2004; Imbens and Wooldridge, 2009; Morgan and Harding, 20006; Nichols, 2007; Becker and Caliendo, 2007; Becker and Ichino, 2002; Nannicini, 2001; Rosenbaum and Rubin, 1983) for the empirical approach used in this study. Empirical applications of these procedures have appeared in the agricultural economics literature (Liu and Lynch, 2007; Mayen et al., 2010; Taur, 2009; Uematsu and Mishra, 2012; Pufahl and Weiss, 2009).

Ideally, the estimate of the ATE would simply be the difference of two outcomes for the same unit; when the unit is assigned to the treatment and when it is not (Imbens and Wooldridge, 2009). In this study, for example, the ATE is considered the difference between crop production costs on certified organic acreage and on conventional acreage. The ATE on the outcome variable in the population of interest can be expressed as:

$$ATE = E(Y_{1i} - Y_{0i}) \tag{1}$$

where Y_{1i} is the outcome variable (production costs) with treatment (organic certification) and $Y0i$ is the outcome variable without treatment. If each individual Y_{1i} and Y_{0i} could be observed among a large representative population, ATE could be estimated by the average value of $Y_{1i} - Y_{0i}$ for the sample of N observations using:

$$ATE = \frac{1}{N}\sum_{i=1}^{N}(Y_{1i} - Y_{0i}) \tag{2}$$

However, with observational data only Y_{1i} or Y_{0i} can be observed because assignment to the treatment is mutually exclusive. Thus, estimating the ATE of organic production on costs is like estimating the counterfactual or imputing missing data. That is, for the treated, it is necessary to estimate the effect of the treatment had the observation not been treated, and for the untreated, to estimate the effect had the observation been treated.

The issue about the evaluation of treatment-effects has triggered a body of research about various estimation techniques. One alternative to deal with potential bias in the estimate of the ATE is to match observations in both the treatment and control groups based on observable characteristics. The propensity-score method uses the predicted probability of being in the treatment group estimated from a binary-choice (logit or probit) model (Rosenbaum and Rubin, 1983). An important feature of the propensity-score model (PSM) is that it summarizes information from multiple variables that may influence choice of the treatment into a single-index variable (Becker and Ichino, 2002). The predicted probability, or propensity-score, of each observation being treated is used to match each treated observation with an untreated observation with a similar propensity-score. Propensity-scores are used to reduce selection-bias by equating groups based on observed covariates.

The propensity-score is defined as the conditional probability of treatment given the observed background variables:

$$p(x) \stackrel{\text{def}}{=} \Pr(T=1 | X=x) \qquad (3)$$

The treatment is (conditionally) unconfounded with the outcome if potential outcomes are independent of the treatment conditional on the observed variables X. This can be written as:

$$Y(0), Y(1) \perp (T|X) \qquad (4)$$

where $\perp$ denotes statistical independence. If unconfoundedness holds, then:

$$Y(0), Y(1) \perp (T|p(X)) \qquad (5)$$

The unconfoundness assumption of the matching estimator implies that any remaining difference in the outcome variable after matching can be solely attributed to the treatment status (Imbens, 2009) and that assignment to the treatment can be considered purely random among matched observations (Becker and Ichino, 2002).[26] This is not a testable assumption, therefore, matching estimators cannot be said to completely eliminate selection-bias due to unobservable factors correlated with both assignment to the treatment and the effect. However, the propensity-score plays an important role in balancing the groups to make them comparable. Rosenbaum and Rubin (1983) showed that treated and untreated subjects with the same propensity-score have identical distributions for other baseline variables. This "balancing property" means that by controlling for the propensity-score when comparing groups, the observational study has been effectively converted into a randomized block experiment where "blocks" are groups of subjects with the same propensities. Assignment of the untreated to the treated is said to be random for observations with the same propensity-score (see Becker and Ichino (2002) for more detail).[27]

Another estimation technique that addresses the potential for self-selection bias uses a parametric model assuming a joint normal distribution between the errors of a selection equation (choice of the treatment) and effect equations. This technique corrects for self-selection bias, due to the correlation of unobserved variables with both choice of the treatment and the effect, allowing for an unbiased estimate of the impact that choice of the treatment has on the effect. In this study, differences in management and input quality are largely unobservable but may be correlated with both the choice of organic crop production and the level of production costs.

In the model with endogenous treatment-effects, the decision to choose the treatment can be expressed with a latent variable indicating the net benefit from choosing the treatment so that:

$$T_i^* = Z_i \gamma + u_i \tag{6}$$

where $T_i = 1$ if $T_i^* > 0$. Z_i is a vector of observable variables expected to affect the choice of the treatment, such as operator, farm, and regional characteristics. The impact of the treatment on the outcome variable can be expressed by:

$$Y_i = X_i \beta + T_i \delta + \varepsilon_i \tag{7}$$

where X_i is a matrix of observable variables expected to impact the outcome variable, such as crop production practices, and farm and operator characteristics, among others.

Equation (7) cannot be estimated directly because the decision to choose the treatment may be determined by unobservable variables that may also be correlated with the effect. If this is the case, the error terms in equations (6) and (7) will be correlated, resulting in a biased estimate of ö. This selection-bias can be accounted for by assuming a joint normal error distribution with the following form:

$$\begin{bmatrix} u \\ \varepsilon \end{bmatrix} \sim N \left(\begin{bmatrix} 0 \\ 0 \end{bmatrix}, \begin{bmatrix} 1 & \rho \\ \rho & \sigma_\varepsilon^2 \end{bmatrix} \right) \tag{8}$$

and by recognizing that the expected effect of choosing the treatment is given by:

$$E(Y_i|T_i = 1) = X_i\beta + \delta + \rho\sigma_\varepsilon\lambda_i \tag{9}$$

where λ_i is the inverse Mills ratio. To derive an unbiased estimate of δ, the two-stage approach begins with a binary model estimation of equation (6). In the second stage, estimates of γ are used to compute the inverse Mills ratio, which is included as an additional term in a least-squares estimation of equation (7). This two-stage Heckman procedure is consistent, albeit not efficient. Efficient maximum likelihood parameter estimates can be obtained by maximizing:

$$L(\gamma,\beta,\sigma,\rho) = \prod_{A_{i=0}} \int_{-\infty}^{0}\int_{-\infty}^{\infty} f(A_i^*, y_i; \gamma, \beta, \sigma, \rho) dy dA^*$$

$$\cdot \prod_{A_{i=1}} \int_{0}^{\infty}\int_{-\infty}^{\infty} f(A_i^*, y_i; \gamma, \beta, \sigma, \rho) dy dA^* \tag{10}$$

where $f(A_i^*, Y_i; \gamma, \beta, \rho)$ is the joint normal density function, which is a function of the parameters. In practice, the negative of the log of the likelihood function is minimized using the estimates from the Heckman procedure as starting values. Once estimated, the difference in costs between the treated and untreated groups, the ATE, is determined by (Greene, 2000, pg. 934):

$$E(Y_i|T_i = 1) - E(Y_i|T_i = 0) = \delta + \rho\sigma_\varepsilon\left[\frac{\phi_i}{\Phi_i(1-\Phi_i)}\right] \tag{11}$$

where φ is the standard normal density function and Φ is the standard normal cumulative distribution function evaluated using the selection equation estimates. This study measures the treatment-effect of certified organic production on costs of producing corn, wheat, and soybeans using these two very different techniques common in the literature to deal with selection-bias. Matching samples using the propensity-score was developed by statisticians as a way of adjusting the sample so that it more closely resembles the results of a randomized experiment (Rosenbaum and Rubin, 1983). In contrast, the approach developed by economists (Heckman) is a way of adjusting the model for potential selection-bias by treating it as a problem of missing variables without adjusting the sample. Treatment-effects are estimated in this study using the Stata software. The "psmatch2" procedure is used to estimate the ATE with the propensity-score matching technique, to assess the common support and matching quality, and to conduct sensitivity analysis. The "etregress" procedure is used to correct for self-selection bias using the technique of linear regression with endogenous treatment-effects. Results of the two treatment-effect estimators are compared with the mean difference in the outcome variable for treated and untreated groups. Empirical estimates of the two treatment-effect models are shown for corn in appendix tables 1 and 2, for wheat in appendix tables 3 and 4, and soybeans in appendix tables 5 and 6. These estimates are used to compute the treatment-effects for the propensity-score matched sample (appendix tables 1, 3, and 5) and the model of linear regression with endogenous treatment-effects (appendix tables 2, 4, and 6) for each crop. For brevity, only the models of linear regression with endogenous treatment effects for total economic costs are shown.

Appendix Table 1. U.S. corn production: Binomial probit estimates to compute propensity scores for the matching analysis of treatment-effects

Variable description	Coefficient	Standard error
Constant	0.4422**	0.2190
Size (100 corn acres harvested)	-0.1322**	0.0311
Size squared	0.0019**	0.0008
Age class (younger than 50 years old)	0.0538	0.1226
Education class (attended college)	0.5104**	0.1246
Primary occupation is off-farm	-0.3505**	0.1708
Location in Lake States[a]	0.1851	0.1398
Location in Northeast States[a]	0.1573	0.2242
Location in Plains States[a]	-0.3548*	0.1998
Continuous row crop rotation	-0.3196**	0.1443
Idle year in crop rotation	0.3269**	0.1599
Moldboard plow used	0.6412**	0.1558
No-till planter used	-1.1297**	0.2304
Irrigation	-0.3713	0.4339
Applied commercial fertilizer	-1.8534**	0.1400
Applied manure or compost	0.2859**	0.1284
Pseudo R^2	0.54	
Likelihood ratio ÷2	685	

[a]Deleted location is Corn Belt.

Note: Dependent variable in the probit equation is whether the farm produced organic corn (0,1). * and ** denote statistical significance at the 10-percent and 5-percent levels, respectively.

Source: USDA, Economic Research Service and USDA, National Agricultural Statistics Service, 2010 Agricultural Resource Management Survey of corn producers.

Appendix Table 2. U.S. corn production: Linear regression estimates with endogenous treatment-effects

Variable description	Coefficient	Standard error
Treatment (organic)		
Constant	-0.3134**	0.0987
Size (100 corn acres harvested)	-0.2120**	0.0248
Size squared	0.0033**	0.0005
Age class (less than 50 years)	0.1721*	0.0927
Education class (attended college)	0.2642**	0.0902
Primary occupation is off-farm	-0.4662**	0.1281
Location in Lake States[a]	-0.3346**	0.1009
Location in Northeast States[a]	0.1185	0.1759
Location in Plains States[a]	-0.8451**	0.1367
Effect (total economic costs)		
Constant	4.4115**	0.6104
Size (100 corn acres harvested)	-0.0788*	0.0449
Size squared	0.0013	0.0012
Location in Lake States[a]	0.4787	0.3080
Location in Northeast States[a]	0.8153	0.5718
Location in Plains States[a]	0.3351	0.3809
Continuous row crop rotation	-0.7646**	0.2805
Idle year in crop rotation	-0.1605	0.3938
Moldboard plow used	-0.1438	0.4175
No-till planter used	-0.1952	0.2678
Irrigation	0.3044	0.5323
Applied commercial fertilizer	0.0827	0.4410
Applied manure or compost	0.1421	0.2857
Organic	1.9178**	0.7912
Sigma	4.1709**	0.0809
Rho	0.0011	0.948
Log likelihood	-4,324	

[a]Deleted location is Corn Belt.

Note: Dependent variable in the probit equation is whether the farm produced organic corn (0,1). * and ** denote statistical significance at the 10-percent and 5-percent levels, respectively.

Source: USDA, Economic Research Service and USDA, National Agricultural Statistics Service, 2010 Agricultural Resource Management Survey of corn producers.

Appendix Table 3. U.S. wheat production: Binomial probit estimates to compute propensity-scores for the matching analysis of treatment-effects

Variable description	Coefficient	Standard error
Constant	0.2619	0.2027
Size (100 wheat acres harvested)	-0.0790**	0.0001
Size squared	0.0004**	0.0001
Age class (younger than 50 years old)	0.3076**	0.1330
Education class (attended college)	0.2941**	0.1303
Primary occupation is off-farm	-0.3168	0.2383
Location in Central States[a]	0.2308	0.1692
Location in Southern Plains States[a]	-0.4317**	0.2134
Location in Northwest States[a]	-0.4259*	0.2541
Continuous row crop/small grain rotation	-0.2878	0.1767
Idle year in crop rotation	-0.3232*	0.1743
Moldboard plow used	0.6940**	0.2164
No-till planter used	-0.6636**	0.1633
Irrigation	-0.4144	0.2804
Applied commercial fertilizer	-1.7940**	0.1329
Applied manure or compost	0.8703**	0.1723
Pseudo R^2	0.52	
Wald ÷2	574	

[a]Deleted location is Northern Plains.

Note: Dependent variable in the probit equation is whether the farm produced organic wheat (0,1). * and ** denote statistical significance at the 10-percent and 5-percent levels, respectively.

Source: USDA, Economic Research Service and USDA, National Agricultural Statistics Service, 2009 Agricultural Resource Management Survey of wheat producers.

Appendix Table 4. U.S. wheat production: Linear regression estimates with endogenous treatment-effects

Variable description	Coefficient	Standard error
Treatment (organic)		
Constant	-0.4932**	0.1161
Size (100 wheat acres harvested)	-0.0986**	0.0115
Size squared	0.0005**	0.0001
Age class (less than 50 years)	0.2101**	0.1011

Appendix Table 4. (Continued)

Variable description	Coefficient	Standard error
Education class (attended college)	0.0891	0.0975
Primary occupation is off-farm	-0.1681	0.1729
Location in Central States[a]	-0.2118*	0.1139
Location in Southern Plains States[a]	-0.4765**	0.1635
Location in Northwest States[a]	-0.8162**	0.1737
Effect (total economic costs)		
Constant	7.9634**	0.6192
Size (100 wheat acres harvested)	-0.0222	0.0224
Size squared	0.0001	0.0002
Location in Central States[a]	-0.7324	0.4628
Location in Southern Plains States[a]	2.9966**	0.5309
Location in Northwest States[a]	1.1475**	0.4531
Continuous row crop/small grains rotation	-0.9350**	0.4588
Idle year in crop rotation	-0.0676	0.4487
Moldboard plow used	-0.0338	0.7068
No-till planter used	-0.3863	0.3434
Irrigation	-0.7153	0.6189
Applied commercial fertilizer	0.0656	0.4508
Applied manure or compost	0.0187	0.6625
Organic	3.9005**	0.9638
Sigma	5.8787**	0.1066
Rho	-0.0028	0.0732
Log likelihood	-5,315	

[a]Deleted location is Northern Plains.

Note: Dependent variable in the probit equation is whether the farm produced organic wheat (0,1). * and ** denote statistical significance at the 10-percent and 5-percent levels, respectively.

Source: USDA, Economic Research Service and USDA, National Agricultural Statistics Service, 2009 Agricultural Resource Management Survey of wheat producers.

Appendix Table 5. U.S. soybean production: Binomial probit estimates to compute propensity-scores for the matching analysis of treatment-effects

Variable description	Coefficient	Standard error
Constant	0.7994**	0.1768
Size (100 soybean acres harvested)	-0.2412**	0.0321
Size squared	0.0038**	0.0007
Age class (younger than 50 years old)	-0.2920	0.1232

Variable description	Coefficient	Standard error
Education class (attended college)	0.1310	0.1163
Primary occupation is off-farm	-0.2243	0.1527
Location in Lake States[a]	0.2454*	0.1358
Location in Plains States[a]	-0.6365**	0.1702
Continuous row crop rotation	-1.6258**	0.1409
Idle year in crop rotation	-0.1805	0.1600
Moldboard plow used	0.9094**	0.1654
No-till planter used	-1.2505**	0.1526
Irrigation	0.5730*	0.3055
Applied commercial fertilizer	-1.1949**	0.1663
Applied manure or compost	0.6481**	0.1666
Applied manure or compost	0.8703**	0.1723
Pseudo R^2	0.55	
Wald $\div^2$	740	

[a]Deleted location is Corn Belt.

Note: Dependent variable in the probit equation is whether the farm produced organic soybeans (0,1). * and ** denote statistical significance at the 10-percent and 5-percent levels, respectively.

Source: USDA, Economic Research Service and USDA, National Agricultural Statistics Service, 2006 Agricultural Resource Management Survey of soybean producers.

Appendix Table 6. U.S. soybean production: Linear regression estimates with endogenous treatment-effects

Variable description	Coefficient	Standard error
Treatment (organic)		
Constant	-0.1469	0.1038
Size (100 soybean acres harvested)	-0.3340**	0.0290
Size squared	0.0051**	0.0006
Age class (less than 50 years)	0.0145	0.0942
Education class (attended college)	0.2631**	0.0901
Primary occupation is off-farm	-0.2354**	0.1153
Location in Lake States[a]	-0.2090**	0.0986
Location in Plains States[a]	-0.5621**	0.1279
Effect (total production costs)		
Constant	8.3588**	0.4905
Size (100 soybean acres harvested)	-0.0886**	0.0450
Size squared	0.0011	0.0014
Location in Lake States[a]	-0.3689	0.3133

Appendix Table 6. (Continued)

Variable description	Coefficient	Standard error
Location in Plains States[a]	0.9594**	0.3698
Continuous row crop rotation	-1.9207**	0.3587
Idle year in crop rotation	-1.5488**	0.4540
Moldboard plow used	1.2219**	0.5239
No-till planter used	0.0730	0.2627
Irrigation	0.1311	0.5267
Applied commercial fertilizer	0.4007	0.2710
Applied manure or compost	-1.0151**	0.4584
Organic	6.9226**	0.6586
Sigma	4.9150**	0.0855
Rho	-0.0364	0.0586
Log likelihood	-5,508	

[a] Deleted location is Corn Belt.

Note: Dependent variable in the logit equation is whether the farm produced organic soybeans (0,1). * and ** denote statistical significance at the 10-percent and 5-percent levels, respectively.

Source: USDA, Economic Research Service and USDA, National Agricultural Statistics Service, 2006 Agricultural Resource Management Survey of soybean producers.

End Notes

[1] Exempting growers selling $5,000 or less a year, who must still comply and submit to a records' audit if requested, but do not have to formally apply.

[2] USDA also supports organic agriculture through the Environmental Quality Incentives Program (EQIP) Organic Initiative, which provides financial assistance to organic producers implementing conservation practices that address a broad array of resource concerns (USDA/Natural Resources Conservation Service (NRCS)).

[3] The relationship between organic crop yields and experience with organic production was evaluated but was not statistically significant.

[4] Food-grade organic crops are generally lower yielding than feed-grade organic crops. Average organic food-grade soybean and wheat yields from the ARMS were not statistically different than average organic feed-grade soybean and wheat yields. Organic food-grade corn yields averaged about 25 bushels per acre less than organic feed-grade corn yields but food-grade corn comprised only about 10 percent of organic corn acreage. Food- and feed-grade organic acreage and production were not delineated in the 2011 Organic Production Survey.

[5] Crop rotations were identified according to what was planted on the field over a 3-year period.

[6] Idled cropland refers to land in cover and soil-improvement crops and cropland on which no crops were planted during a growing season. Some cropland is idle each year for various

physical and economic reasons, such as to promote the accumulation of soil moisture and to enhance soil fertility and organic matter.

[7] Organic producers may or may not have used organic seed, depending upon its availability. The regulatory text says, "Non-organically produced seeds and planting stock that have been treated with a substance included on the National List of synthetic substances allowed for use in organic crop production may be used to produce an organic crop when an equivalent organically produced or untreated variety is not commercially available."

[8] Organic producers had higher capital costs because they used more field operations, particularly for tillage. As more capital assets such as tractors are used in production, the greater the capital consumption and the annual capital charge.

[9] Organic feed-grade corn comprised 90 percent of organic corn sales and received lower prices than food-grade corn. The average price received for organic feed-grade corn was $6.96 per bushel compared with $7.92 per bushel for organic food-grade corn. Production cost differences between organic food- and feed-grade corn were not statistically significant.

[10] Organic prices for corn, wheat, and soybeans were those reported as received by farmers. The harvest-period price for conventional corn, wheat, and soybeans was that received by growers during the most active harvest month of each crop. Many conventional crop producers store grain with the expectation that higher prices in future months will more than cover the additional costs of storage, hauling, and marketing, so many conventional producers may have received a higher price than the mean harvest-period price reported here. The harvest-period price is used to value crop production because the additional costs of crop storage and marketing are not included in the accounts.

[11] Winter wheat was the predominate type of wheat produced by both conventional and organic growers, accounting for 69 percent of conventional acreage and 58 percent of organic acreage. Durum wheat was produced on only 5 percent of conventional and 6 percent of organic wheat acreage, while the remainder, 26 percent of conventional and 36 percent of organic acreage, was planted with other spring wheat.

[12] Some organic wheat and soybean producers used seed saved from the previous crop. The cost of this homegrown seed was determined by the opportunity cost of using this seed, defined by using organic crop prices at the previous crop harvest to value the amount of seed planted.

[13] The coefficient of variation (CV) on the estimates of organic wheat operating and operating plus capital costs was 29 and 13 percent, respectively, compared to CVs less than 2 percent for conventional wheat cost estimates.

[14] Organic food-grade wheat comprised 89 percent of organic wheat sales and received higher prices than feed-grade wheat. The average price received for organic food-grade wheat was $9.77 per bushel compared with $7.33 per bushel for organic feed-grade wheat. Production cost differences between organic food- and feed-grade wheat were not statistically significant.

[15] In recent years, the growing problem of herbicide-tolerant weeds has caused some conventional corn and soybean producers to increase the use of mechanical weed control with a row cultivator, among other measures. (For more on this topic, see Gunsolus, 2013.)

[16] Costs of food-grade and feed-grade organic soybeans were not significantly different. The average price received for food-grade soybeans ($15.08 per bushel) was significantly higher than that for feed-grade soybeans ($12.48 per bushel).

[17] The matching was done with replacement, meaning that conventional producers could be matched with more than one organic producer. Based on visual inspection of the propensity-score density, 10 percent of the data were trimmed from each tail of the distribution for each

crop in order to improve the data overlap and common support (Caliendo and Kopeing, 2008).

[18] Exogenous variables are those that influence choice of the treatment without being affected by it. For example, operator age is exogenous to choice of the organic approach, while crop rotation is not because the crop rotation is affected by whether or not the organic approach was chosen.

[19] Results of the linear regression with endogenous treatment-effects models for total economic costs are shown for the sake of brevity. Models for operating costs and operating plus capital costs yielded similar results.

[20] Robust standard errors were estimated for both treatment-effect models. The method derived by Abadie and Imbens (2014) was used to estimate standard errors for the propensity-score matching estimator.

[21] Delate et al. (2006) point out that for crops such as corn and soybeans, where most of the production is from transgenic crops, increased interest in nontransgenic food ingredients has created markets where producers may obtain a price premium for crops produced during organic transition years.

[22] This assumed planning horizon is arbitrary, but reasonable given the average age of the organic producers of each crop.

[23] This procedure assumes that the same crop—corn, wheat, or soybeans—was produced on the land during each year of the transition period. In actuality, a typical organic rotation, including forage crops and/or an idled year, would likely be included during the transition period. This assumption was necessary because only data on the additional costs of producing the target crop were known, and may possibly indicate a higher charge during the transition period than if the costs of rotated crops were reported.

[24] Possible source of selection-bias are the differences between organic and conventional producers in unobserved variables, such as the level of management and input quality. Users of organic systems may exhibit a higher level of management, which would be correlated with both the treatment and effect. Also, selection-bias could result if organic crops are planted on higher quality land than conventional crops. These unobserved variables could result in mean organic costs appearing to be lower than they would otherwise, and more similar to mean conventional costs. The treatment-effect models address this issue.

[25] The organic corn average price premium estimated from the 2010 ARMS was $2.71 per bushel in the Lake States (Upper Midwest) and $2.12 per bushel in the Northeast, compared with $3.08 per bushel in other States. However, the average total costs of organic production were much lower in the Lake States and the Northeast, resulting in net returns that were very similar among the regions.

[26] This assumption is termed in various ways, such as "ignorability" (Wooldridge, 2010, p. 908), "selection on observations" (Fitzgerald et al., 1998), and "unconfoundedness" (Imbens, 2004; Rosenbaum and Rubin, 1983; Wooldridge, 2010, p. 908).

[27] It has been argued that matching models are special cases of selection models which assume that conditioning on observable variables eliminates self-selection bias (Heckman and Navarro-Lozano, 2004; Mayen et al., 2010). That is, matching models create the conditions of an experiment in which the treatment variable is randomly assigned. However, the matching model does not directly account for correlation among unobservable variables that could bias the treatment-effect. Imbens (2004) suggests that the assumption about the distribution of unobserved variables being similar for treated and untreated agents is ultimately an empirical question.

In: Profitability of Organic Field Crops
Editor: Madeline Rose Bowers
ISBN: 978-1-63484-167-2
© 2015 Nova Science Publishers, Inc.

Chapter 2

ORGANIC FARMING SYSTEMS[*]

Catherine Green and Robert Ebel

The organic label is the most prominent food eco-label in the United States. In 2000, USDA published national organic standards that reflected decades of private-sector development. USDA's national regulatory program is designed to facilitate interstate trade, reduce consumer fraud, and provide consumer assurance that all organic products sold in the United States meet a high national standard. All organic growers, processors, and distributors are required to meet the national standard and be certified by a USDA-accredited State or private group unless they sell less than $5,000 annually in organic products.

USDA regulations define organic farming as an ecological production system that fosters resource cycling, promotes ecological balance, and conserves biodiversity. Organic farmers are required to avoid most synthetic chemicals and must adopt practices that maintain or improve soil conditions and minimize erosion. Organic production systems can be used to increase farm income, as well as reduce pesticide residues in water and food, reduce nutrient pollution, improve soil tilth and organic matter, lower energy use, reduce greenhouse emissions, and enhance biodiversity (U.S. Department of Health and Human Services, 2010; Greene et al., 2009; Ribaudo et al., 2008).

---

[*] This is an edited, reformatted and augmented version of a chapter that originally appeared in Economic Information Bulletin No. 98, Agricultural Resources and Environmental Indicators, 2012, issued by the U.S. Department of Agriculture, Economic Research Service, August 2012.

In 2005, USDA began to include targeted oversamples of organic producers in its Agricultural Resource Management Survey (ARMS), which collects detailed information about farmers' production practices, as well as costs and returns in major farm sectors. Some of the differences in practices and characteristics of organic and conventional production systems are apparent from survey responses by soybean, wheat, apple, and corn producers (fig. 3.5.1).

CONSUMER DEMAND DRIVES GROWTH IN THE ORGANIC SECTOR

Organic food sales in the United States have increased from approximately $11 billion in 2004 to an estimated $25 billion in 2011 (fig. 3.5.2). Market penetration has also grown steadily; organic food products accounted for more than 3.5 percent of total U.S. food sales in 2011. Although the annual growth rate for organic food sales fell from the double-digit range in 2008 as the U.S. economy slowed, it still far outpaces the annual growth rate in all food sales (*Nutrition Business Journal*, 2012).

ADOPTION OF ORGANIC SYSTEMS IS HIGHEST FOR SPECIALTY CROPS

U.S. producers dedicated approximately 4.8 million acres of farmland— 2.7 million acres of cropland and 2.1 million acres of rangeland and pasture— to organic production systems in 2008 (latest year for which data are available). Top States for certified organic cropland include California, Wisconsin, North Dakota, Texas, and Minnesota (fig. 3.5.3). Top States for certified organic pasture and rangeland are Wyoming, New Mexico, Texas, California, and South Dakota. Overall, the adoption of organic farming systems is low—only about 0.7 percent of all U.S. cropland and 0.5 percent of all U.S. pasture was certified organic in 2008.

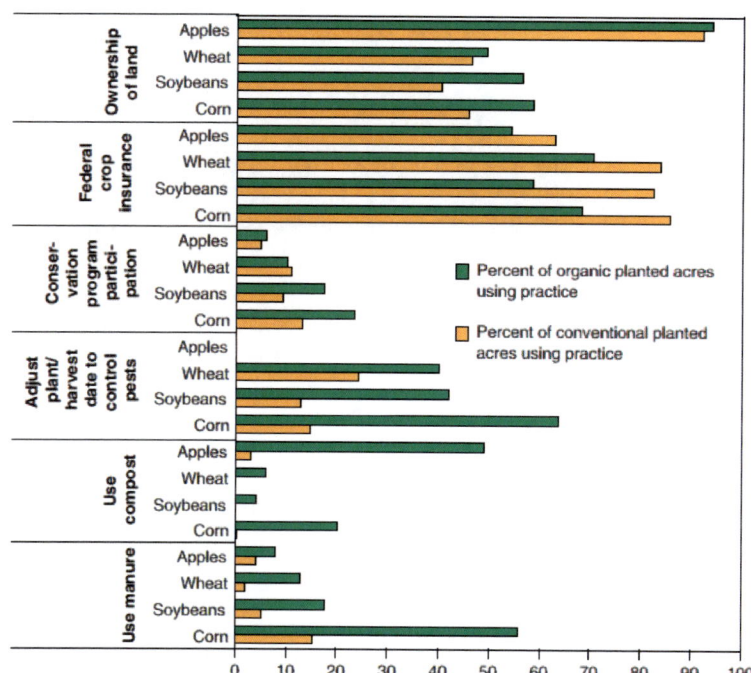

Source: USDA, Economic Research Service and National Agricultural Statistics Service, based on data from the Agricultural Resource Management Survey.

Figure 3.5.1. U.S. organic and conventional operations: selected characteristics and practices.

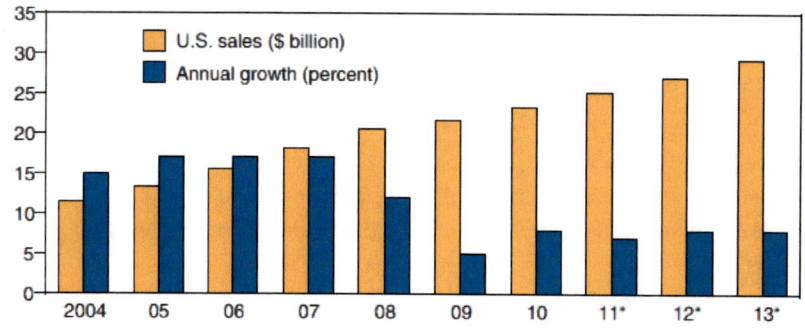

* 2011-13 estimates or projections. Source: Nutrition Business Journal.

Figure 3.5.2. Organic food sales in the United States, 2004-2013.

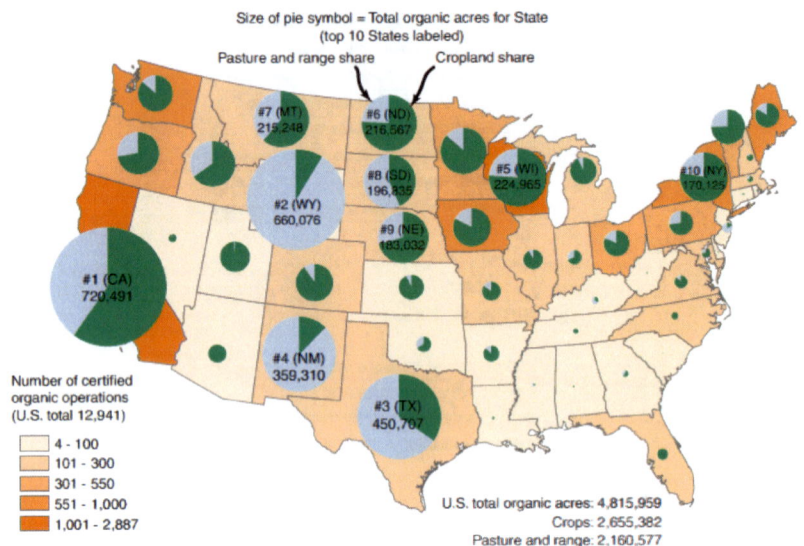

Source: USDA, Economic Research Service, based on information from USDA-accredited certifiers.

Figure 3.5.3. Organic operations accounted for less than 1 percent of total crop acreage in 2008.

Still, many U.S. producers are embracing organic farming in order to lower input costs, conserve nonrenewable resources, capture high-value markets, and boost farm income. While only a small percentage of the top U.S. field crops—corn (0.2 percent), soybeans (0.2 percent), and wheat (0.7 percent)—were certified organic in 2008, nearly 9 percent of U.S. vegetable crop acres and 3 percent of U.S. fruit and tree nut acres were grown under certified organic farming systems. Markets for organic vegetables, fruits, and herbs have been developing for decades in the United States, and fresh produce is still the top-selling organic category in retail sales. Organic livestock was beginning to catch up with produce in 2008, with 2.7 percent of U.S. dairy cows and 1.5 percent of layer hens managed under certified organic systems.

Obstacles to adoption by farmers include high managerial costs and risks of shifting to a new way of farming and limited knowledge of organic farming and marketing systems. According to Lynn Clarkson, a top organic grain broker, specific obstacles to adoption in organic grain production include the 3-year lag due to the organic transition period requirement, fewer organic marketing outlets, the need for onfarm storage, the lack of third-party contractors for organic pest and nutrient management, heavy managerial requirements, fear of criticism from neigh-bors, unknown risks, lack of government infrastructure support, and subsidies for ethanol that increase demand for conventional grain supplies (Clarkson, 2007).

Producers also face many challenges once they have shifted to organic production. Respondents to USDA's 2005 organic dairy producer survey indicated that certification paperwork and compliance costs were the most challenging aspect of organic milk production, followed by finding new organic sources of feed and dairy replacements, higher costs of production, and maintaining animal health (McBride and Greene, 2009). In the produce sector, a recent California study of small and mid-sized organic farmers, producing mostly fruit and vegetables, found that more than 80 percent reported marketing challenges—having too much or too little volume, obtaining organic price premiums, locating and accessing markets, handling competition, and accessing information on pricing (Cantor and Strochlic, 2009).

REFERENCES

Cantor, Alida, and Ron Strochlic. 2009. Breaking Down Market Barriers for Small and Mid-Sized Organic Growers. California Institute for Rural Studies. Nov.

Clarkson, Lynn. 2007. "Review of economic impacts of production, processing, and marketing of organic agricultural products." Statement of the President of Clarkson Grain Co., Inc., Subcommittee on Horticulture and Organic Agriculture—Public Hearing. 110th Congress 2007-2008, Witness Opening Statements, House Committee on Agriculture, April 18.

Greene, C., C. Dimitri, B. Lin, W. McBride, L. Oberholtzer, and T. Smith. 2009. Emerging Issues in the U.S. Organic Industry. EIB-55. U.S. Department of Agriculture, Economic Research Service, June.

McBride, W.D., and C. Greene. 2009. Characteristics, Costs, and Issues for Organic Dairy Farming. ERR-82. Nov.

Nutrition Business Journal (NBJ). 2012. U.S. organic food sales – chart 22. Penton Media, Inc.

Ribaudo, M., L. Hansen, D Hellerstein, and C. Greene. 2008. The Use of Markets To Increase Private Investment in Environmental Stewardship. ERR-64. U.S. Department of Agriculture, Economic Research Service, Sept.

U.S. Department of Agriculture, Agricultural Marketing Service. 2000. "National Organic Program; Final Rule, 7 CFR Part 205," Federal Register, Dec. 21, www.usda.gov/nop.

U.S. Department of Agriculture, Economic Research Service. 2009. Data sets: Organic Production. http://www.ers.usda.gov/data-products/organic-production.aspx.

U.S. Department of Agriculture, Economic Research Service. 2011. Data sets: ARMS Farm Financial and Crop Production Practices. http://www.ers.usda.gov/data-products/arms-farm-financial-and-cropproduction-practices.aspx.

U.S. Department of Health and Human Services, National Institutes of Health, National Cancer Institute.

2010. Reducing Environmental Cancer Risk: What We Can Do Now, President's Cancer Panel, 2008-09 Annual Report, April.

For more information, see...

U.S. Department of Agriculture, Economic Research Service. 2012. "Organic Agriculture." http://www.ers.usda.gov/topics/natural-resources-environment/organic-agriculture.aspx

INDEX

A

accounting, 4, 8, 53
actuality, 54
age, 17, 24, 54
agencies, 5, 6
agricultural economics, 42
agriculture, 2, 40, 52, 60
alfalfa, 20, 25
assessment, 31
assets, 8, 18, 53
audit, 52
average total costs, 54

B

basic research, 14
bias, 7, 37, 38, 43, 44, 45, 46, 54
biodiversity, 5, 55

C

capital consumption, 53
carbon, 9
causal inference, 38
causality, 4
certification, vii, 1, 3, 4, 5, 6, 8, 10, 15, 18, 28, 30, 31, 32, 33, 34, 35, 36, 37, 42, 43, 59

CFR, 60
challenges, 59
chemical(s), 2, 4, 9, 18, 20, 22, 25, 37, 55
coefficient of variation, 53
commercial, 2, 3, 7, 16, 19, 20, 22, 23, 26, 47, 48, 49, 50, 51, 52
commodity, 8, 28, 42
competition, 4, 59
compliance, 59
composition, 22, 25
compost, 19, 20, 22, 23, 26, 47, 48, 49, 50, 51, 52
composting, 4
computing, 28
conditioning, 54
Congress, 59
conservation, 52
control group, 43
correlation, 44, 54
cost, 3, 4, 6, 8, 14, 18, 20, 23, 26, 27, 28, 29, 30, 31, 32, 33, 34, 35, 36, 38, 53
costs of production, 7, 29, 32, 59
criticism, 59
crop(s), vii, 1, 2, 3, 4, 5, 6, 7, 8, 9, 10, 14, 15, 16, 17, 19, 20, 22, 23, 25, 26, 27, 28, 29, 30, 31, 32, 33, 35, 36, 37, 38, 40, 42, 43, 45, 47, 48, 49, 50, 51, 52, 53, 54, 58
crop producers, 3, 7, 9, 15, 17, 37, 38, 53
crop production, vii, 1, 2, 3, 4, 7, 9, 10, 15, 16, 33, 36, 42, 43, 45, 53
crop residue, 4

crop rotations, 5
cumulative distribution function, 46
cycling, 4, 5, 55

D

dairy industry, 37
Department of Agriculture, 41
Department of Health and Human Services, 55, 60
dependent variable, 28
depreciation, 18
disclosure, 20, 23, 26
distribution, 4, 24, 45, 53, 54
drought, 14

E

Economic Research Service (ERS), 1, 6, 8, 10, 20, 21, 23, 24, 27, 29, 32, 33, 36, 39, 42, 48, 49, 50, 51, 52, 55, 57, 58, 59, 60
education, 15, 24, 28
electricity, 18
employment, 17, 24
energy, 55
environment, 4
environmental impact, 14, 15
erosion, 55
ERS, 5, 6, 18, 42
ethanol, 59
expertise, 15

F

farmers, 4, 5, 37, 53, 55, 56, 58, 59
farmland, 56
farms, 2, 3, 7, 8, 14, 16, 17, 19, 20, 21, 22, 25, 28, 37
fear, 59
Federal Register, 5, 39, 60
fertility, 37, 53
fertilizers, 20, 22
field crops, vii, 1, 2, 5, 7, 10, 27, 38, 58
field trials, 2

financial, 8, 14, 18, 52, 60
financial performance, 14
fluctuations, 11
food, 5, 11, 12, 13, 24, 34, 35, 37, 41, 52, 53, 54, 55, 56, 57
forage crops, 54
fraud, 55
fruits, 58
funding, 7
funds, 6

G

greenhouse, 55
groundwater, 4
growth, vii, 1, 2, 5, 6, 56
growth rate, 56

H

harvesting, 4
health, 9, 59
herbicide, 24, 40, 53
high school, 18, 19, 22, 24, 26
House, 59
human, 15

I

imports, 6
income, 8, 55, 58
independence, 44
independent variable, 28
infrastructure, 59
ingredients, 37, 54
interest rates, 15
investment, 31, 32
Iowa, 8, 38, 39
irrigation, 18
issues, 42

L

labeling, 5
lead, 15
light, 38
livestock, 58
lower prices, 53

M

machinery, 25
major issues, 9
management, 4, 14, 15, 37, 38, 40, 45, 54, 59
manure, 4, 19, 20, 22, 23, 26, 47, 48, 49, 50, 51, 52
marketing, 53, 58, 59
matrix, 45
measurements, 14
media, 41
Mexico, 56
Missouri, 8
models, 28, 30, 36, 38, 47, 54
moisture, 53
Montana, 22
motivation, 15

N

National Institutes of Health, 60
Natural Resources Conservation Service (NRCS), 42, 52
nitrogen, 4
normal distribution, 44
nutrient, 4, 15, 55, 59

O

obstacles, 58
Oklahoma, 8, 21
operating costs, 3, 18, 20, 22, 23, 25, 28, 30, 31, 32, 54
operations, 2, 5, 8, 18, 19, 20, 22, 23, 24, 25, 26, 30, 31, 53, 57, 58
opportunities, 14
opportunity costs, 18
organic food, 4, 5, 11, 34, 37, 52, 53, 56, 59
Organic Foods Production Act, 5
organic matter, 53, 55
organism, 4
overlap, 54

P

parameter estimates, 46
pasture, 5, 56
pesticide, 55
pests, 37
pollution, 4, 55
population, 8, 36, 43
President, 59, 60
prevention, 4
probability, 43, 44
producers, 2, 4, 7, 8, 9, 10, 15, 16, 17, 18, 20, 21, 22, 23, 24, 25, 27, 28, 29, 30, 32, 34, 36, 37, 48, 49, 50, 51, 52, 53, 54, 56, 58
production costs, 4, 7, 8, 10, 14, 15, 16, 18, 22, 28, 29, 30, 32, 33, 34, 37, 38, 43, 45
profit, 2, 37
profitability, vii, 1, 3, 4, 10, 15
property taxes, 18

R

rangeland, 56
real estate, 18
recession, 6
recovery, 18, 31
regression, 7, 27, 28, 29, 30, 31, 33, 47, 48, 49, 51, 54
regression model, 28, 30
regulations, 5, 55
regulatory framework, 5
reimburse, 6
replication, 14

requirement(s), 5, 10, 58, 59
researchers, 7
residues, 55
resources, 3, 5, 15, 18, 58, 60
restrictions, 5
retail, 58
risks, 58
rotations, 52

S

school, 18
science, 42
Secretary of Agriculture, 5
seed, 2, 9, 16, 18, 19, 20, 22, 23, 24, 26, 37, 53
sensitivity, 47
software, 47
South Dakota, 8, 18, 22, 56
soybeans, 2, 3, 5, 6, 7, 8, 12, 16, 20, 22, 25, 29, 30, 31, 32, 33, 34, 35, 36, 37, 46, 51, 52, 53, 54, 58
soymilk, 25
specialty crop, 7
species, 4
standard error, 54
state, 41
stock, 53
storage, 37, 53, 58

T

target, 8, 54
target population, 8
Task Force, 38
techniques, 4, 43, 46
tofu, 25
total product, 51
trade, 55

transition period, 31, 54, 58
treatment, 4, 7, 8, 15, 18, 27, 28, 29, 30, 31, 32, 33, 34, 35, 36, 38, 42, 43, 44, 45, 46, 47, 48, 49, 50, 51, 54

U

U.S. Department of Agriculture (USDA), 1, 5, 6, 7, 8, 10, 11, 12, 13, 16, 17, 18, 20, 21, 23, 24, 27, 29, 30, 32, 33, 36, 39, 41, 42, 48, 49, 50, 51, 52, 55, 56, 57, 58, 59, 60
U.S. economy, 56
United States, vii, 1, 5, 6, 7, 38, 40, 55, 56, 57, 58

V

variable costs, 3, 14
variables, 4, 28, 30, 43, 44, 45, 46, 54
variations, 37
varieties, 2, 9, 16
vector, 45
vegetables, 58, 59

W

wage rate, 15
Washington, 8
water, 18, 55
waterways, 4
Wisconsin, 8, 18, 39, 56

Y

yield, 16, 35, 36